T0130773

THE MATHEMATICS OF THE GODS AND THE ALGORITHMS OF MEN

ALSO BY PAOLO ZELLINI

A Brief History of Infinity

THE MATHEMATICS OF THE GODS AND THE ALGORITHMS OF MEN

A CULTURAL HISTORY

PAOLO ZELLINI

TRANSLATED BY SIMON CARNELL AND ERICA SEGRE

PEGASUS BOOKS
NEW YORK LONDON

THE MATHEMATICS OF THE GODS AND THE ALGORITHMS OF MEN

Pegasus Books Ltd.
148 W. 37th Street, 13th Floor
New York, NY 10018

First Pegasus Books edition May 2020

Library of Congress Cataloging-in-Publication Data is available.

ISBN: 978-1-64313-414-7

10 9 8 7 6 5 4 3 2 1

Printed in the United States of America
Distributed by Simon & Schuster

Contents

Introduction 3

1 Abstraction, Existence and Reality 11
2 Mathematics of the Gods 20
3 Mathematical and Philosophical Formulas 36
4 Growth and Decrease, Number and Nature 43
5 *Katà gnómonos phýsin*: The Nature of the Gnomon 55
6 *Dýnamis*: The Capacity to Produce 61
7 Intermission: Spiritual Mechanics 72
8 Zeno's Paradoxes: The Explanation of Movement 76
9 The Paradoxes of Plurality 89
10 The Limited and the Limitless: Incommensurability and Algorithms 97
11 The Reality of Numbers: Cantor's Fundamental Sequences 114
12 The Reality of Numbers: Dedekind's Sections 125
13 Mathematics: A Discovery or an Invention? 139
14 From the Continuum to the Digital 144
15 The Growth of Numbers 157
16 The Growth of Matrices 165

17 The Crisis of Fundamentals and the Growth of Complexity: Reality and Efficiency 183

18 *Verum et Factum* 192

19 Recursion and Invariability 196

Notes 201

Index 229

Introduction

About which reality does mathematics speak to us? It is widely supposed that mathematicians preoccupy themselves with abstract formulas, and that it is only for inexplicable reasons that these formulas have applications in every area of science.

We conceive of immaterial entities that seem subsequently to be destined to define models of phenomena that actually occur in the world. On the one hand, there are real, present things; on the other, mathematical concepts, creations of our mind which simulate their behaviour in a more or less effective way. Ignorance of the true reason behind the descriptive power of formulas and equations certainly doesn't help to clarify the underlying motivation behind mathematical thinking. It gives currency instead to the idea that mathematicians are not inclined to engage with the world. Mathematics continues to present itself as a science that elaborates ingenious operations with rules and concepts that seem to have been conjured up with the sole aim of their being executed correctly.[1] It matters little that some of these ideas have been suggested by the observation of natural phenomena; the operations rapidly produce advanced and complex concepts that distance themselves from observable reality and ultimately confirm the distorted image of mathematics as a merely linguistic game or a series of empty formulas.

If we turn to its remote history and to its deepest purpose, however, mathematics appears to be orientated very differently than is commonly supposed. Its origins allow us to understand that ancient arithmetic and geometry were beginning to assume the role not so much of describing or simulating real things as offering a foundation for the very reality of which they were a part. It was concrete things themselves – those that were directly and immediately perceivable – that were shifting and mutable, and therefore prone to appear unreal. To find precisely what removed them from such instability and evanescence, one had to look instead to numbers, to their relations and to the figures of geometry.

If we think about Zeno's famous paradoxes, the number-points of the Pythagoreans and the atomists of antiquity, about Plato's mathematical philosophy, the discovery of incommensurability and the significance of the concept of relationality (*lógos*), about Babylonian calculus and Vedic mathematics, we are faced with a great mass of knowledge designed to capture the most internal and invisible – as well as the most real – aspect of the things that exist in nature. The theory of numbers and of continuous mathematics elaborated in the nineteenth century presented itself as the ideal continuation of ancient Pythagoreanism, and of a vision of the world inspired by an intuition of the atomistic nature of reality. Mathematicians of the period, then, continued to contend that their symbolic constructions corresponded to very real entities, and the widespread impression was that on the success of their theories depended the foundation of knowledge that was necessary for understanding the world. When in the early twentieth century the principles of these theories became uncertain and began to undergo a critical revision, mathematics was obliged to search for the reasons that make a system of calculation genuinely concrete and reliable.

At this time a key term began to circulate persistently among mathematicians: *algorithm*, a word which denoted not so much an abstract formula as an actual process.[2] This process needed to unfold in a finite number of steps, from an initial set of data to a final outcome in space and time, according to the modalities predicted by a machine. The formal definition of 'algorithm', based on recursion, on Turing's machine or on other formulations, dates back to the fourth decade of the last century – but the first indications that it would be this concept of algorithm that would inherit the sense of mathematical reality – that is to say, all that mathematicians consider to be real and actual – were already being witnessed in the first decade of the twentieth century, in the early signs of mathematical intuitionism and in the first arguments with which the French mathematician Émile Borel confronted the semantic paradoxes and incipient crisis of fundamentals.

The science of algorithms followed a tumultuous arc of development that spanned the entire century, reaching a culmination in the formal definitions of the thirties, before bifurcating into two different but complementary trends after the construction of the first digital calculators: on the one hand, information theory, with its abstract notions of computability and of computational complexity; on the other, a science of calculus on a large scale, dedicated to resolving mathematical problems in physics, economics, engineering and computer science in purely arithmetical and numerical terms. The multiple philosophical facets of this second trend have not yet been sufficiently analysed, but it is already evident to all how much it has contributed in every area of life, to culture and social organization, with the multiplication of a diverse variety of calculation processes aimed at solving specific problems of the most varied kind.

In numerical calculation on a large scale the theoretical *effectiveness* of algorithms aims at achieving computational *efficiency*. And today it seems clear that, in order to be *real*, the very same mathematical entities that have been constructed through a process of calculation must be capable of being thought of in the same way as efficient algorithms. Today the efficiency depends above all on the way in which they facilitate growth in computational complexity and errors of calculation. In particular, the error depends on *how rapidly the numbers grow* in the course of calculation.

The reasons for the growth in the numbers are strictly mathematical and may be analysed thanks to relatively advanced theorems. But it is worth noting that the reason for this growth, in all of its aspects, was already the object of the closest scrutiny in ancient thought, and it is precisely the way in which the growth of quantities is treated in Greek geometry, in Vedic calculations and in Mesopotamian arithmetic that has contributed to the understanding of the causes of the growth of numbers in modern algorithms. The reason for this is as simple as it is surprising: some important *computational schemas* have remained unchanged since then, right up until the most complex strategies of which large-scale calculation avails itself today.

Where do these schemas derive from? In certain cases of particular relevance to modern science, the sources are clear: these schemas derive from a singular combination of human design and divine dictate. In Vedic India the altars of Agni, the fire god, had complex geometrical forms and needed to be capable of being enlarged a hundredfold without changing shape, by using specific techniques that can also be found in Greek geometry and Mesopotamian calculation. In Greece it so happened, as in the case of doubling the cube, that the enlargement of a form was also demanded by a deity. But the enlargement of the geometric form was in strict relation to the algorithms delegated to approximate those numbers which, when faced with having to measure geometric magnitudes such as the diagonal of a square, or the relation between a circumference and diameter, are irrational. It was the Vedic gods and those of Greece, long before the God of Descartes, who guaranteed a nexus between mysticism and nature, between our most intimate sphere and external reality. Mathematics offered at that stage, too, the principle of this possible connection. At any rate, the modalities of growth in ancient geometry, inspired by religious observance, are reflected today in the growth of numbers in digital calculations, having an essential impact on the stability of calculation itself and on the predictive power of mathematical models. In fact, the modalities of growth of *geometric* figures, in particular the square, are often correlated to *numerical* procedures that generate fractions p/q, which approximate irrational numbers, where p and q are whole numbers. But usually p and q grow much more rapidly the more rapid the convergence in method, with potential negative effects on the precision and stability of the entire process of calculation.

The thesis that irrational numbers are real entities, with an ontological status comparable to that of whole numbers, was an achievement of the mathematics of the end of the nineteenth century, and of the way in which the concept of the arithmetical continuum was defined at the time. But the development of the science of algorithms and digital calculation became in the twentieth century the expression of a new kind of opposition: a kind of final act of the perennial tension – already touched on in Zeno's paradoxes – between numbers and geometry, between the discrete and the continuous. It is legitimate to speak of a clash between them because the study of algorithms was propitiated from the very beginning of the twentieth century by a contest of ideas aimed at re-evaluating the most realistic and constructive aspects of mathematics, in contradistinction to those abstractions which had given rise to paradoxes and a crisis of fundamentals: on the one hand, the masterly command of Émile Borel which signalled the importance of defining mathematical entities through algorithmic constructions; on the other, the dramatic schism articulated by L. E. J. Brouwer and by mathematical intuitionism within the compendium of mathematics. Arguing that a number exists only if it is built, Brouwer issued a general challenge to the prevailing scientific system, calling into question the fundamental definitions of classical analysis.

The constructivist philosophies, based on the idea of actual calculability, have assigned new pre-eminence to that which seemed alien to the abstract vocation of mathematics; they have given importance, that is, to the concrete operation, the factuality of nature and, ultimately, to the computational process that develops inside a machine operating within the limits allowed by space and time. But it is equally evident that important computational strategies are modelled on the same schemas that mankind had elaborated during those eras when they were closely engaged in supposed communications with the gods. For ritual purposes, in Vedic India as in Greece, the reason for growth in magnitude was of fundamental importance and had to be confronted mathematically. And the basic blueprints for enlarging a geometric shape are still traceable in the most advanced computational mathematics. The schemas have not changed, though they have certainly been elaborated and perfected through complex mathematical theories. From these theories we also derive the rationale for their efficiency and their effective capacity to translate mathematical models of nature into pure digital information.

The same computational process, the same process articulated in a myriad of concrete automatic operations, can take place only thanks to abstract mathematical structures, inserted more or less artificially into the calculation. Mathematical abstraction is combined in a necessary and systematic way with the materiality of the automatic execution of operations. The calculation is made possible because of complex theoretical presuppositions and the special properties of numbers, functions and matrices.

The question thus remains open: are numbers real entities? And if we were to answer in the affirmative, are they all 'real' in exactly the same way? The two questions need to be tackled together. The history of the last century and an analysis of the concepts of number and algorithm allow us to see our way towards an initial conclusion: different kinds of numbers exist that do not have the same ontological status, but in relation to which it is possible to ascribe an existence in reality, for a variety of reasons and from different points of view. A key criterion for establishing the reality of numbers is the way in which they grow in the processes of calculation. And the first reason for this phenomenon should be sought in the analysis of geometrical growth elaborated in ancient thought, especially in Greek, Vedic and Mesopotamian mathematics.

1. Abstraction, Existence and Reality

Where does mathematics come from, and what are its objectives? Why are there triangles, squares, circles and pentagons? What kind of reality and existence may be attributed to numbers? Mathematics, as even some of the most intransigent formalists often admit, is real knowledge – and the object of this knowledge, we can say with confidence, is not arbitrary, does not depend on a capricious imagination or on the arbitrary choice of certain axioms or principles. Furthermore, it frequently happens to be perceived as an external reality, independent of the mind that elaborates it.

We usually think that mathematics is an abstract science, since in practice it extracts relationships and patterns from specific entities such as numbers in order to study common properties – as if the properties were in turn new entities that obey their own laws, with the advantage that all that may be said about the latter may be applied to the different specific entities from which the abstraction has been made. The reasoning, when it is abstract, thus becomes more general and more powerful. But abstraction makes more problematic the identification of an *essence* of mathematical entities, the location of an intrinsic character susceptible to definition. From their existence, at least as objects of our thought, a real essence of something that is stable and clearly recognizable does not seem to derive. The traditional reciprocal connection between *essentia* and *existentia* – whereby one term would not achieve the status of the real if the other were absent – seems to have been lost.[1] Does the essence of numbers consist in some kind of special nature that pertains to them which we can apprehend directly, or does it derive rather from the properties of an abstract domain of which the numbers in question are only a possible – not necessarily unique – example?

It is usually the second idea that is favoured. The properties that are studied separately form a system of truths derived from axioms – and mathematics, dissociated from any form of direct intuition, would then be considered as a science of pure formal relations, independent of every concrete interpretation of them. A specific numerical field,[2] such as that of real or complex numbers, satisfies the same axioms to which *other* mathematical fields also conform. It can thus happen that a number field may be identified only by an *isomorphism*, because it is indistinguishable from other mathematical entities that have the same properties. It's a circumstance that is enough to make mathematics completely impervious to the possible search for the specific nature of those numbers.

The recognition, or characterization, of a mathematical object usually depends on its specification in relation to properties that are independent of any possible construction or representation. And it is sometimes sufficient for a single, simple definition based on a few properties to identify an entire class of isomorphic domains.[3]

Are the instruments of logic suitable for establishing whether, and under what circumstances, a mathematical entity has real existence in some form or other, in an external world that is independent of us? There are conflicting views on this matter. There is, for instance, a significant difference between the views of Gottlob Frege and Bertrand Russell on this subject, to cite two of the pre-eminent exponents of *logicism*, which inspired the idea which sought to demonstrate that all mathematics is reducible to logic. For Frege, numbers were logical objects that one must define in some way. They are not created through definition: the definition merely demonstrates what exists in its own right.[4] Russell's position is different – decidedly more inclined towards *nominalism*, but not rigidly so. For Russell, logic as a whole is an indispensable instrument of knowledge of the *external* world. In *Our Knowledge of the External World* (1914), he wrote that the fundamental elements for explaining the nature of events are things, qualities and *relationships*, previously ignored by logic, which had assumed that all statements should have a subject–predicate form. Traditional logic had not taken into account the *reality* of relations, which were in fact indispensable to descriptions of the world and were needed in order to dissipate the errors of traditional metaphysics. In all likelihood, according to Russell, it was precisely traditional mysticism and metaphysics which had posited the idea of the unreality of the world that we perceive. Logic, Russell goes on to explain, serves to articulate a description of the world that originates with atomic propositions, which register the facts of empirical experience. From the atomic propositions we move to more complex ones, always thanks to logic. Logic obviously does not enter into the registration of elementary facts, but constitutes a comprehensive understanding, *a priori* in character, on which all potential deductions are based. This complex of deductions, while not deriving exclusively from sensory knowledge, should be considered to be knowledge that is real and effective. And mathematics, Russell concludes, is an essential part of it.

Nevertheless, if we pay attention to what logic actually tells us, at stages what seems to prevail is the most radical nominalism. Russell could not avoid acknowledging that the categories which define numbers are not things of which we have to establish the existence. Of course, with the language of logic, one strives to declare that a number, a set or a function exists. But the *reality* of numbers does not coincide at all with the existence established by this language. The logical propositions which assert that a mathematical entity exists serve 'not so as to succeed in knowing that which exists, but so as to know what a given assertion or given thesis, belonging to us or to others, *says* that there is; and this is properly speaking a linguistic problem, not an ontological one'.[5]

Logic alone does not suffice to establish an ontology of abstract objects. Nor should we be astonished by the words of Goodman and Quine, summarizing in 1947 the radical nominalism that was implicit in the attempt to base mathematics on logic:

> We do not believe in abstract objects. Nobody supposes that abstract entities – classes, relations, propositions, etc. – exist in space-time; but we want to say something more than this. We want to do away with their existence altogether.[6]

It is still worth specifying that 'abstract objects' are susceptible to existing in a variety of forms, and to becoming embodied in entities that are relatively concrete, with an existence in space and time. This circumstance depends on at least two distinct and different factors: the existence of an automatic calculation that develops in the physical space and time of a machine, and the widespread conviction that mathematical entities resemble living organisms, to the extent of being able to dictate the concrete conditions which permit us to study and understand them. The possibility of operating, subject to certain conditions, on numbers that exist physically as binary sequences in the memory of a calculator represents in itself a decisive fact. John von Neumann was able to observe how automatic calculation on a large scale, which had been progressing since the forties with the development of the first digital calculators, consisted of a calculation in *space* and *time* that was made possible by abstract structures that only pure mathematics was capable of elaborating. Mathematical abstraction and physical reality were inseparable in the new science of calculus.

When we talk about 'reality' we are straying into an extremely arduous terrain where things may be upended: a number may also be real that does not exist as we would wish or expect but that we somehow feel obliged to define in the course of meeting our needs or those of the algorithm.[7] In any case, Quine himself, in defining real numbers, was referring to preceding theories such as those of Dedekind and Weierstrass, which were in turn indebted to the various sources of knowledge stretching back *at least* to the theory of proportions in Book V of Euclid's *Elements*. And this theory was enhanced, in its own turn, by a knowledge of computation that was much more ancient than Euclidean theories. That which appears to be decisive in gauging the degree of reality is a range of experiences and situations, both theoretical and practical, that often have their origins in the remote past and which make it almost essential to arrive at a certain definition or posit a specific theory, that wants it to be configured in a certain way and not in another. Ernst Mach wrote passages of great interest on the way that scientific discoveries occur: one always ends up revisiting that which was already known, in an uninterrupted combination of the unknown and the known, of analysis and synthesis, of discovery and recognition. Discovery and invention are usually sustained by a preceding history, and by a chain of ideas that have facilitated them and rendered them necessary.

To simplify things, one might begin by assuming that a *realist* is someone who contends that the propositions of a scientific theory are either true or false, and that which makes them so is something external to ourselves, something different from sensory data, from our language and our thought.[8] So might it be plausible, then, to advance the notion that the reality of numbers depends on that of the physical world? The countless applications of mathematics to physics, economics, engineering, chemistry and information technology form a body of knowledge of impressive breadth and efficacy that seems capable of establishing the reality of formulas according to some kind of naturalistic philosophy, based on the idea that the universe consists exclusively of natural objects – objects situated in space-time and subject to causal laws. The universe is a vast book, wrote Galileo, in a famous passage in *The Assayer*, written in the language of mathematics and impossible to understand without resorting to triangles, circles and other types of geometrical form. But it is doubtful that this circumstance alone gives reality to mathematical entities; we can only affirm that it is not possible to be a realist with regard to physical theory and a nominalist with regard to the theory of mathematics.[9]

Therefore, it is not a given that a philosophical realism regarding physics is the only way of establishing a mathematical reality. However much it is our mind that elaborates the calculations, it is widely recognized that, if we put aside their applications to the real world, we find ourselves ultimately faced with algorithms, formulas and demonstrations that have more or less variable contours but are capable of dictating regardless of us the conditions and modalities of their existence, like wilful and obstinate creatures. This is why mathematicians so frequently claim that they feel obliged to recognize certain entities and not others, to be genuinely astonished that they seem to come from another world, or at least from a reality which is outside of our perceptual capacity, language and mental framework[10] – a circumstance which is enough in itself to make us reconsider Kant's thesis on the *a priori* character of mathematical thinking, according to which 'reason perceives only that which it produces itself according to its own design' (*Critique of Pure Reason*, Preface to the second edition, 1787). Mathematical entities, as conceived by many scientists and as they were perceived to be by Charles Hermite in particular, are not lifeless, artificial constructions but real, living beings with their own kind of coherence and intentionality, capable of guiding our research and determining the conditions of that freedom and autonomy that we customarily attribute to the dictates of our own reason.

'To define reality,' noted Simone Weil, 'nothing is more important than that.' And her own first conclusion was that: 'The real is transcendent; and this is Plato's essential idea.'[11] Hence it is necessary, for at least two reasons, to begin with the mathematics of antiquity: the question of the reality of mathematical entities has been around since the time of Pythagorean philosophy as well as Vedic and Mesopotamian calculations – and modern calculus, from the sixteenth century to the present day, has been based on constructions developed from ancient science. The question of the reality of numbers was posed in ancient Greece, at least implicitly, in attempts to understand what relations mathematical entities have with the infinite, with the non-being of the *ápeiron*, or infinite. Infinity was absence (*stéresis*), pure potentiality – and everything, to exist and to endure, had to oppose itself against the negativity of the limitless. This was, in Greek mathematics, the task of the *lógos*, of proportion, in which the precursors of modern numbers were found. The phenomenon of relation, and what derived from it, was an entity close to the gods. And not only in Greece. It was precisely in this *relation* that one needed to find the reality of numbers: not coincidentally, it was thanks to the reprising of concepts of relation and proportion (found in Euclid and in pre-Euclidean computational science) that the mathematicians of the nineteenth century established the ontology of numbers and the arithmetical continuum. It was this same circumstance that gave value to the logic of quantifiers through which we can affirm that a certain class *exists* (the existential quantifier), or that the conditions which allow a number to be identified must be satisfied for *every* value of certain variables (the universal quantifier) – a circumstance which allowed Bertrand Russell to assert that 'the sense of reality is vital in logic'.[12] This was not enough to really establish an ontology, but what was done subsequently to remedy such inadequacy is an ideal continuation of computational processes already well known in remote eras – in China, India and Mesopotamia as well as in Greece.

The processes of enumeration employed since the most ancient times often had the purpose of making things – both men and gods – real by means of a demonstrative or divine gesture, introducing them on to the stage of the world in order to render them actual and recognizable, with a full right to exist in space and time. Enumerations, censuses, lists and catalogues recur frequently in Homer and Hesiod, in Aeschylus and Herodotus, as well as in the Old Testament. Enumeration was a prerogative of the *lógos*, which suggested a process of selection and collection, of aggregation ordered by means of different entities into a single whole. Numbers themselves, in the Pythagorean tradition, were separated like a set of points in space, with an order similar to that of armies or of stars in the heavens. Consequently, the numbers with which one counted were not only abstract but also real and manifest. Nevertheless, with the demonstration that incommensurable quantities exist it was discovered that the natural number, the *arithmós*, was not enough to give actual existence to those entities that today we call irrational numbers. There are relations between magnitudes which cannot be expressed as relations between whole numbers. As Georg Cantor would go on to demonstrate in the nineteenth century with the *diagonal method*, phantasmal entities exist that escape from any kind of enumeration. Greek mathematics attempted to represent these entities indirectly, by means of a sequence of numbers, relations and geometrical figures regulated by a law of growth and decay. Modern mathematics would go on to use analogous sequences to define irrational numbers.

2. Mathematics of the Gods

It is hard to say why and where mathematics first arose. It is also pointless, perhaps, if we listen to the arguments of those who have sought to discourage any search for its origins by showing the intrinsic senselessness of such attempts. In *Human, All Too Human* (par. 249), Nietzsche denounces the weighing down, the deep weariness that can result from a gaze cast systematically towards the past. In the second of his *Untimely Meditations*, 'On the Uses and Disadvantages of History for Life', he argues that the soul of antiquarian man, who guards and archives the past in the most exacting way, can fall victim to the blind frenzy of a collecting mania and to an oppressive curiosity that acts as an obstacle to any impulse towards all that is new and vital. After Nietzsche, Michel Foucault explains that the origins of any system of knowledge are numerous and that it makes no sense to seek only one, as if it were the sole source of everything.

It seems to me that mathematics is hardly exempt from this rule: in any case, there are many kinds of mathematics – arithmetic, mathematical physics, algebra, geometry, analysis and statistics – that provide answers to different questions, obey a diversity of criteria and use various techniques of demonstration, even if in many cases the study of abstract mathematical structures has revealed surprising affinities between disparate domains, allowing us to glimpse the outlines of a unified knowledge. Over the course of time mathematical theories have often changed aspect, have been conceived and formulated in many ways and with various outcomes, branching out into other theories.

Mathematics has been born and reborn countless times, so much so in fact as to lead one to think of a concept of tradition like that in the fable of Moses and Rabbi Akiva in the Talmud.[1] Moses, according to the fable, receives the Torah from God written in a script ornamented with infinite flourishes and crowns. For every flourish, God predicts, an as yet unborn person called Akiva will one day formulate innumerable doctrines. Moses asks to be enlightened, and God allows him, as if by magic, to sit alongside the pupils in the eighth row of the schoolroom in which Akiva is about to teach. Bewildered, he understands nothing of what Akiva is explaining. But he is reassured when the pupils ask the rabbi how he has taken the right route to tackle a certain question – and he replies that it was by way of a law given to Moses on Mount Sinai. This contains a deep truth about tradition, about the continuous readapting of ancient knowledge in the light of more advanced formulations that in turn become new forms of understanding. Mathematics is no exception to this process.

We can, however, identify some foundational themes that explain how and why mathematics took the form that it did, over time, and not some other one instead. At the end of the nineteenth century it seemed finally to be clear from which presuppositions one should start. Mathematics, it was theorized, is born out of the free activity of the intellect, from a *spirit* of calculation unconditioned by extraneous circumstances, and by means of simple set operations. The simplest intellectual operation, fundamental and apparently immune to contradiction, became that of grouping together different entities (numbers, functions, matrices) into a single class and attempting to operate with these entities, if possible, without leaving that class. Mathematicians were seeking to demonstrate how the class operated in is *closed*,[2] a reassuring property, because it prevents the possibility of having to contend, in the course of a calculation, with an alien entity that does not comply with the rules established for that class. Sometimes you find yourself obliged to leave a certain designated class, as in the case of solutions to an algebraic equation which can have real or complex numbers and consequently require a departure from the field of rational numbers. As is well known, the extension to complex numbers emerged from research conducted on the resolution of third-degree algebraic equations. Rafael Bombelli, following in the wake of Girolamo Cardano, was its principal architect. But the field of complex numbers is itself closed with regard to operations of addition and multiplication that generalize, within the field, the ordinary operations between rational or real numbers.

The study of ever more abstract domains did not undermine the conviction, gradually reached in the nineteenth century, that everything may be traced back to the simple concept of the whole number, while the whole number may in turn be based on the even simpler concept of the set, or set theory. The great constructions of Weierstrass, Peano, Cantor, Dedekind or Veronese were all inspired by the grand project of *arithmetizing* analysis, of making the entire mathematical edifice derive, at least in principle, from whole numbers, from set theory and from the concept of limit.

It is necessary, however, to distinguish between *origin* and *derivation*. Did mathematics really originate from the operations of set theory such as those elaborated by nineteenth-century science? If we adhere to the sources, then the answer cannot be other than negative.

In ancient Mesopotamian mathematics we find arithmetical calculations similar in kind to modern procedures: genuine algorithms *avant la lettre*. Vedic geometry relating to the construction of fire altars includes the study of equivalence between geometric figures, such as the circling of the square, and to achieve this requires accurate numerical processes, such as the approximation of $\sqrt{2}$. For the Pythagoreans, numbers had a geometrical form, a circumstance which by definition demanded analysis of the relation between numbers and shapes and which would later lead – we may conjecture – to the discovery of incommensurable quantities. In Greece we find a geometry virtually connected to algebra and to the *computatio* which precedes the modern science of calculus. Its fundamental outlines can also be seen in ancient Chinese mathematics. Therefore, in general, one sees how, in different traditions, an enigmatic relation was established between number and geometrical figure – a problematic tension that in addition to giving rise to the study of incommensurability led to analysis of the concept of the infinite and of the structure of the continuum, as well as to the mathematical *lógos*, a science of relations as the foundation of an intelligible cosmos. And it was from this problematic tension that the first attempts were launched to give shape to important concepts such as incommensurability, effective constructability and approximation which would characterize all subsequent mathematics.

It is impossible to deny that calculation reveals, in its earliest manifestations in remote civilizations, a profound affinity with a variety of practices and systems of knowledge. However scanty the sources may be, one can legitimately argue that mathematics and philosophy, geometry and religion, calculus and metaphysics descended from one great, original, reciprocally combined origin. This admixture in no way compromises the specificity of mathematical knowledge; it does not imply that mathematics serves some other purpose or is in need of external justifications. If anything, the opposite is the case: algebra's formulas and geometry's constructions enjoy an unquestionable specificity, so conclusively evidenced and of such clarity, so to speak, that they do not need to resort to any explanations or justifications outside of themselves. Not even logic is able to give an exhaustive explanation. Calculations appear dry and alien to the philosophical or religious domain, but this is a misleading impression: religion and mathematics, metaphysics and calculus, ritual action and exacting thought seem to combine, in the beginning, into a single, powerful nexus – a combination that one must seek to discern in the great designs of ancient cosmology, in the conceptions of the first philosophers, as well as in the computational strategies and most minute calculations of which Greek, Chinese and Babylonian mathematicians were so fond.

From the sources we can deduce that the first to pose mathematical problems were not men but gods – or at any rate men inspired by the gods. This is confirmed by an impressive variety of stories, explanations and concordances. The Prometheus of Aeschylus is an inventor of numbers and the first to distinguish the astronomical signs that regulate the great temporal cycles, and this would have required a mathematical theory of relations. Analogous prerogatives were attributed to the Egyptian god Thoth; to the alter ego of Prometheus, Hermes; and not least to Palamedes, the pupil of the centaur Chiron and hero of the Trojan War. Plato alludes to him in the *Phaedrus* (261 b–d), calls him 'Eleatic' (an adherent of the philosophical school of Elea) and attributes to him logical skills equal to those of Zeno. To Gorgias we owe the *Defence of Palamedes* (82 B 11a DK), which forms part of an epic tradition alternative to that of Homer in which Odysseus is cast in a somewhat less exalted light. As Iamblichus asserts in *The Pythagorean Life* (31), of the three beings with the gift of reason one is god, the other is man and the third has the character of Pythagoras. Pythagoras himself learned from the Egyptians and Babylonians an art of measurement transmitted from the gods. Parmenides discovered from a goddess that being is singular and perfect. His metaphysical vision, compatible with a mathematical conception of the world, could be aligned with the idea of absolute space that Newton would delineate in his *Principia.* Plato asserts that mathematics was granted to us by Uranus, the god of the heavens, who expressed the laws of numbers and of relations through the movements of the stars (*Epinomis*, 977b).

We know that the gods of the Vedic tradition had a prerogative: to be utterly clear and truthful. And this, as Charles Malamoud notes, was also their weakness, necessitating an extreme caution and diffidence in their dealings with mankind. Transparency and veracity implied exactitude; hence the altars dedicated to them needed to be accurately and precisely designed and constructed, as did the performance of ritual actions and ceremonies. But this very veracity might lead, for self-protection, to a systematic concealment: the names of the deities were deliberately garbled in order to make them inaccessible, thus intensifying their mystery and truth. It is precisely with this distortion of the name, according to the *Śatapatha Brāhmana*, that mysticism originates. We are reminded in this regard of the famous words spoken by Hesiod's Muses (*Theogony*, 27–8): 'We know how to utter many lies that resemble the truth, but we also know how, when we choose, to sing that truth.'

Perhaps this was one of the things that made mathematics so appealing, for here was an exact science that proposes *enigmas*, abstruse problems to be unlocked by way of stratagems based on calculations, demonstrations and algorithms. Zeno's paradoxes concerning the impossibility of motion are among the most celebrated enigmas. But abstruseness and the enigmatical are still in accordance with the Pythagorean principle that 'falsehood is inimical to the nature of number', as a fragment of Philolaus (44 B 11 DK) has it. The simplest geometrical figures, such as the straight line, the square, the triangle or the circle, configured the most abstruse enigmas in the clearest and most accessible form, such as the incommensurability of quantities or the existence of infinite or infinitesimal quantities. For the Pythagoreans, the definition of shapes had a theological character, as Proclus reminds us in his *Commentary on the First Book of Euclid's 'Elements'* (130), and it isn't at all surprising if each shape was associated with a deity, a custom that would be transmitted down through the centuries, in successive revisitations, at least until the time of Giordano Bruno.

In Greece the resolution of a mathematical problem was translated into a votive offering to the gods, as attested by a letter from Eratosthenes to King Ptolemy which has come down to us via Eutocius of Ascalon, the late commentator on Archimedes: his mechanism for the construction of the proportional mean to duplicate a cube became a votive offering placed within the deity's shrine. This constituted, in Greek, the *anáthema*, a term related to *thésis* (denoting the action of placing or establishing something, also in a judicial context). Now mathematical theorems do not aim for anything other than to demonstrate a *thesis*, and the invention of rigorous demonstration itself could have been the result (following in the wake of a prodigious insight by Simone Weil) of intensified attention directed at images conceived to be reflections or embodiments of divine reality.

What is it that the gods have bequeathed to us? Does mathematics owe them a debt of gratitude for a simple collection of anecdotes and legends, or for a legacy of genuine knowledge? The answer is beyond doubt. From problems posed by the gods stems the basis of a way of thinking without which modern mathematics would be inconceivable.

We can identify the reason behind this persistent link between divine and human orders, between the foundational propositions of the gods and modern mathematics, not so much in general outlines, in explicit ends or in a shared programme of abstraction from the visible world. The connection is more subtle than that, and all the stronger for involving the most technical and secret operations of calculation, the fundamental paradigms on which algebra and analysis are still based today. The decisive link occurs with respect to two main themes: the modality of increase of geometrical figures, and the equivalence of figures of different shape.

The essential references can be found in Vedic treatises such as the *Śulvasūtra* (*śulva* is the rope, the tool, together with pegs, for ritual measurement) on the construction of fire altars dedicated to Agni. These are also tracts designed to clarify the origin and significance of procedures on which Greek mathematics was built. The most important versions of these tracts belong to the school of Baudhāyana, of Āpastamba and of Kātyāyana, and these sources respond above all to the need to establish equivalence between altars of different shapes. The search for equivalence leads to the construction of a square of the same area as a rectangle, or of a circle of the same area as a square. From the circling of a square emerged, presumably, the related and more famous problem of squaring the circle.

The other requirement was to enlarge an altar while keeping its shape unchanged. Whoever was in charge of the ritual had to envisage, at least in principle, the construction in five layers of bricks of altars of increasing size with surfaces equal to 7½, 8½, 9½, up to 101½ *purusha* squared, where *purusha* was a unit of length equivalent to the height of a man with his arms raised. This entailed the solving of various geometrical problems, such as the enlargement of a square, or of finding a square equal to the sum or to the difference between two given squares, problems the resolution of which implied knowledge of Pythagoras' theorem: the square made on the diagonal of a rectangle is equal to the sum of the squares made on its sides.

Vedic geometry is based on geometrical constructions which are described, in a subsequent era, in Euclid's *Elements*. But the two treatments do not always share the same aims: in the *Elements* we find the rigour of demonstration, and in the *Śulvasūtra* what predominates is the idea of a dynamic growth of geometrical figures and the principle of the invariance of shapes when undergoing a change in scale. The distance between Vedic and Greek geometry manifests itself especially in the expression of the formula for the square of a binomial, $(a + b)^2 = a^2 + 2ab + b^2$, an algebraic equivalence of central importance for the development of modern analysis. In the *Śulvasūtra* of Āpastamba (III, 8)[3] certain formulas are given for the enlargement of squares of particular size: a rope of 1½ *purusha* leads to a square with an area equal to 2¼ *purusa* squared, because the formula of the square of the binomial becomes $(1 + \frac{1}{2})^2 = 1 + 1 + \frac{1}{4} = 2\frac{1}{4}$. In other words, if you increase the length of the side 1 by ½, the area of the square increases by 1¼. In a similar way, if to the square of side 2 you add ½, you end up with a square of an area of 6¼ because $(2 + \frac{1}{2})^2 = 4 + 2 + \frac{1}{4} = 6\frac{1}{4}$. The text explains immediately afterwards the more general rule: the increment $2ah + h^2$ to be added to the square of side a, when this undergoes the increase h, is made up of two rectangles, $2ah$, and a square h^2 that together form a gnomon,[4] the squared figure which, when applied to the square of side a, produces a square of greater area $(a + h)^2$. This is the nub of the idea: George Thibaut, who in the nineteenth century provided a translation of and commentary on the Vedic treatises on the construction of fire altars, when summarizing it remarked that Āpastamba and Kātyāyana advanced 'significant examples that illustrate the way in which the increase or decrease in the length of a side causes the increase or diminution of the square'.[5] But in the *Śulvasūtra* we also find the *general rule* of increase and of reduction.

Something that should not be overlooked is the fact that Euclid demonstrates a purely geometrical version of the formula regarding the square of the binomial (*Elements*, II, 4). The rigour manifested in Euclid's demonstration is lacking in the Vedic treatises, but Euclid does not mention the *incremental nature* of the formula, which is the *incipit* not just of the incremental nature of formulas of analysis (from the sixteenth century onwards) but also of the computational procedures with which they are linked. This incremental nature is plainly explicit in the *Śulvasūtra*.

These and other geometrical constructions, such as the cube of the binomial for the enlargement of a geometric cube, have constituted the principal tool for the analysis of algebraic equations – in the Renaissance as in the modern era. In the sixteenth century the discovery of complex numbers, due largely to the work of Rafael Bombelli, derived from the study of second- or third-degree equations based on the rule of (gnomonic) incrementation of a square or a cube. The solving of algebraic equations by means of gnomons proved to be *the* problem *par excellence* – confronted for the first time, in the West, by Theon of Alexandria (an editor of Euclid's *Elements*) in the fourth century, and in subsequent centuries by Arab and Italian mathematicians. Towards the end of the sixteenth century the French mathematician François Viète developed a general method for solving numerically an equation of an arbitrary degree consisting of a kind of algebraic extension of the rule of enlargement of a square or a cube. Newton simplified Viète's procedure, and at the end of the seventeenth century Joseph Raphson discovered in turn how to express Newton's method through a simple iterative formula.[6] In this very same iterative formula we find today the essential model for solving, with automatic procedures of great efficiency, general systems of equations and problems of minimum functions.

But it's not only single numerical procedures that depend on the rules for enlarging a geometrical square. In those same rules we find the foundation of forms of reasoning that gave rise to the development of analysis from the seventeenth century onwards, as well as, in more recent years, to the stabilizing of systems of calculus. As a general rule, it is of utmost importance to know how a function or an entire process of calculation behaves for small variations in or increments of the variables. There is instability every time that *small* variations, increments or disturbances of the variables cause a *big* variation in the value of the function, or in the result of a process of calculation.

To define the principal operations of differential calculus, the algebraic automatism that permitted the manipulation of the new symbols of infinity, Leibniz resorted to formulas such as that for the square of a binomial.[7] And only a few years prior to this, Bonaventura Cavalieri had used the same formulas for his method of the indivisibles (for determining the size of geometrical shapes).[8] Some fundamental strategies for discretizing problems on the continuum, for translating differential or integral problems in which the variables assume the values of the numerical continuum into problems of an arithmetical nature that may be solved with the ordinary operations of sum and product, are based on formulas which generalize the expression of the square of the binomial. To a large extent, the relation between the continuous and the discrete depends on these very same formulas.

It isn't easy to find in the *Śulvasūtra* a precise religious meaning for the geometrical forms of the altars dedicated to Agni, nor is it possible to clarify the issue on the basis of any such treatise, because mathematical thought would go on to adopt, in the centuries that followed, the form with which we are familiar and not another form. One can have recourse, however, to a much vaster literature. Precise allusions and lists of figures, dimensions and geometrical operations occur in texts that are usually assigned origins more ancient than those of the *Śulvasūtra* and which include extensive exegeses of a mythic, ritualistic and metaphysical kind. The *Taittirīya Samhitā* (5, 4, 11) contains a long inventory of altars of various geometrical shape, and quite a few passages in the *Śatapatha Brāhmana* refer to constructions of altars in the shape of a falcon, prefiguring symbolic flight. The geometric forms are measured in *purusha*, and in its most primitive version the stylized body of the altar resembling a bird was composed of seven squares, four of which make up the central trunk, with two for wings and one for a tail. It is written that the *ṛṣi*, the vital spirits, created seven different persons in the form of squares, and that the squares formed a single body, the body with which, in the beginning, Prajāpati, the creator of all things, was made: the creator of the universe and of all mortal as well as immortal beings (*Śatapatha Brāhmana*, VI, 1, 1–3; X, 2, 3, 18).

For what purpose are fire altars built? One starts with the legend of Prajāpati, who, as related in the *Śatapatha Brāhmana* (VI, 1, 1, 12–13), remains shattered and disarticulated as a result of the very creation for which he is responsible. The life force escapes from him and the gods abandon him. But Prajāpati asks Agni to put him back together again, and Agni is precisely the altar that consists of the reconstructed body of Prajāpati.

Figure 1

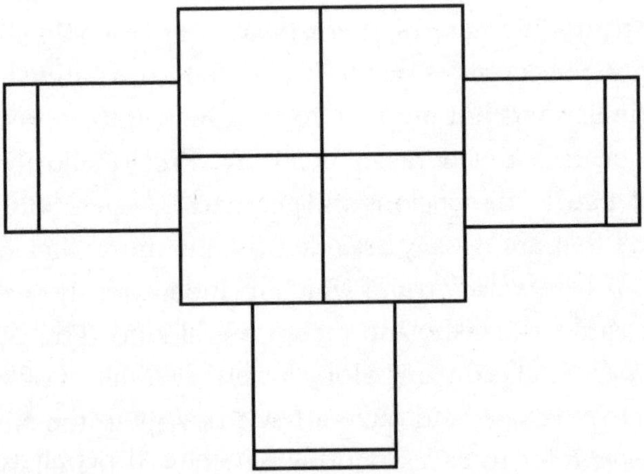

The *Śatapatha Brāhmana* (VI, 1, 2, 36) explains that, assuming the form of a bird, the life-giving spirits became Prajāpati, and that 'assuming that form [of the altar], Prajāpati created the gods'. The nascent gods in turn became immortal. A passage from the *Brhadāranyaka Upanishad* (I, 4, 6) explains that Prajāpati designed the deities that are superior to him and that, despite being himself of merely mortal nature, he generated that which is immortal in nature.

But how could this kind of hypercreation of an immortal happen through the work of a mortal, and how can we conceive of the creator of the universe as a mortal being? From the text of the *Upanishads* and the commentary of Śankara it can be deduced that Prajāpati can be understood from different angles and assumes apparently contradictory characteristics. He situates himself equally both within and beyond transmigratory becoming: 'Although remaining sitting, he was far away; though at rest, he moves everywhere' (Śankara, *Brhadāranyaka Upanishad*, I, 4, 6). If immersed in the becoming of which he is the author, Prajāpati appears mortal, divided, dismembered, prey to fear and loneliness and, consequently, as related in the *Brhadāranyaka Upanishad* (I, 4, 1), needing to counter evil, to *burn away all evil.* This is why he is called Purusha; in fact, Śankara comments, the evil of attachment and ignorance impedes the realization of the divine nature that is proper to Prajāpati. Interpreting symbolically the component parts of the word, Śankara points out that Purusha designates he who was the first (*pūrvam*) to burn (*aushat*) all the evils that obstructed the realization of his true nature, and that the same supplicant, following Prajāpati's reconstruction and the correct celebration of the rite, obtains perfect identification with Purusha (Śankara, *Brhadāranyaka Upanishad*, I, 4, 1).

The construction of the fire altar requires the application of precise norms, mathematical in kind, that we find in the *Śulvasūtra.* But the bricks which properly connected attain the required shapes, for example that of a circle or a square, are themselves akin to gods. Every brick represents a deity that must be invoked, declares the *Katha Upanishad* (I, 1, 15), and Mṛtyu, or Death, which is cognizant of the tools for achieving divine nature and freeing oneself of the fear caused by the manifestation of becoming in the world, shows how many bricks should be used and what the other specifications are for erecting the altar. As in the case of Prajāpati, so Mṛtyu is located both inside and outside the process of becoming: in the former location it is synonymous with death, hunger and fear; in the latter it represents a redemptive deity giving instruction on the power of fire as a means of purification and transcendence.

The altar unites internal and external: Purusha is hidden deeply in all beings, but we do not perceive him because our sensory organs project us outwards to the material world and render us incapable of a properly inward perspective.[9] Agni constructed in a geometrical form is the synthesis of the interior and exterior, of mind and nature – a mathematical solution to a polarity that modern science will experience, in a dramatic way, in the fundamental intellectual crises of the early twentieth century.

A passage of the *Śatapatha Brāhmana* (VI, 1, 3, 20) helps us to better understand what the character of the fire altar is in relation to an idea of numbers and of geometric form subject to a process of growth. It is specified there that the altar must be raised in the space of *one* year and noted that there are those who prefer to determine that it should be erected instead in *two.* The building work and the ritual recitation must last a year, during which the altar is like a seed destined to germinate, 'because the seed that is planted is productive; it lies beneath, changing and growing'. But the growth of the altar must respect precise mathematical proportions: a requirement that accords perfectly with the Platonic vision of numerical progression as a modality of growth that conforms with *phýsis*. In the Platonic vision, such growth started from the number 1, conceived as the generating principle of the progression that is seemingly imprinted upon nature (*phýsis*).[10]

With presuppositions such as this, important procedures of calculus were developed, part of a comprehensive patrimony of mathematical knowledge in which we can still discern the principal devices for the reconstruction and growth of Prajāpati.

We may ask ourselves, finally: what was it that made mathematics the most fitting instrument for the reconstruction of Prajāpati? We can hazard a hypothesis, recalling a passage from a treatise on arithmetic by Boethius. Everything that exists in nature, Boethius explained, owes its form to the laws of mathematics. The connections through which the four elements are combined, the alternation of temporal rhythms, the celestial revolutions: everything can be described by numbers. The altar had an analogous cosmological significance. But the number itself, however subject to variation, because different kinds of numbers exist, always retains the same substance since it is *never composed of entities different from itself*.[11] In the notion of number we find the principle of immutability, the ability to assume all possible forms without ever becoming estranged from oneself. This, precisely, was the nature of the god that needed to be reassembled.

3. Mathematical and Philosophical Formulas

Plato assigned to mathematics a dianoetic/discursive function, that is to say, an intermediate position between the *dóxa* (thought or opinion corrected by and yet distinct from science) and *noûs*, or 'intellection' (the supreme intelligence of philosopher kings (*Republic*, 511 d)). From this stem formulas, expressions and ways of thinking that have marked the entire course of metaphysics as well as mathematics itself. Among the terms they shared were *lógos* (discourse, but also mathematical relations) and *spermatikòs lógos*, the seminal reason of the Stoics, the cosmic principle of generation – but also the unit from which, by means of ingenious algorithms, numerical relations resembling irrational numbers were extracted. *Logistiké* indicated the arithmetical calculation of relations but was also the term used to refer to the faculty of deliberation, similar to calculation, embedded in the eternal part of the soul. The philosophical expressions 'more and less' and 'great and small' occur frequently in Plato's Dialogues as synonyms for the indefinite or infinite (*ápeiron*) but are also adapted to signify the way of approximating irrational numbers with fractions: as one advances with the calculation, the distance between the fraction and the irrational number (not representable as a fraction) becomes increasingly small, while the numerator and the denominator of the fraction become ever larger. Around the middle of the fifteenth century, the Neo-Platonist Nicholas of Cusa declared that truth does not permit of either the more or the less. His words need to be read *in speculo mathematico*: truth is in a central position, like the pivot of an infinitely oscillating balance scale that allows one to get closer to the point of equilibrium with ever more finessed distribution of weights – but without ever reaching perfect equilibrium.

The life of the soul had a mathematical form. As it seems legitimate to deduce from the *Nicomachean Ethics* of Aristotle, the more and the less, the excesses (rounding up) and defects (rounding down) of the mathematical approximations to an irrational number were the counterpart of the incessant oscillations that mark our moral life – between the excesses and defects that place themselves, like Scylla and Charybdis, either side of that middle way in which perfection resides.

The list of philosophical formulas borrowed from mathematics can be extended: *antanaíresis* is the method of searching for the divider between two quantities (the Euclidean algorithm); but the Stoics employed a similar term to denote far-seeing wisdom. The same term, *antanaíresis*, cognate with the German *aufheben*, the Latin *tollere*, with which Hegel would go on to unite the two complementary moments of dialectic, the taking away and the keeping, could also be rendered as 'to resolve', to untangle retrospectively, returning to first principles.[1] So, too, Euclid's *prós ti* (*Elements*, V) expresses the mathematical idea of 'relation', while in Thomas Aquinas the same idea becomes the key to grasping the meaning of the theological concept of creation. One of the assumptions of Platonic doctrine consists of the structural analogy, or homology, between laws – essentially identical and coherent – operating in diverse phenomenal contexts.[2] The same presupposition was shared before Plato by the Orphics and the Pythagoreans.

These laws seem to depend not on the exercise of our judgement but on a kind of necessity intrinsic to the mathematical entities themselves. This is why the formulas seem to be charged with hermetic power: they have a life of their own and are self-perpetuating, in contrast to our own expectations, determining categorically the shape of our reasoning. They seem to belong to an *external reality* outside of our mind, and perhaps this was the reason why Plato would not admit into his Academy anyone who did not know geometry.

Nevertheless, over time the same epistemological formula, the veracity of which was reinforced and justified by affinity with the mathematical formula, demanded its own autonomy, manifesting the ambition to dictate by itself the irrevocable laws of thought. There emerged from this an embarrassment that can be found in the pages of modern scientists such as Galileo, for whom ignorance of mathematics would make any sort of investigation of the natural world impossible. Galileo knew full well that Plato and the Pythagoreans admired the human intellect to such a degree as to consider it as 'partaking of the divine' (*Dialogue Concerning the Two Chief World Systems*, I) on account of its ability to understand the significance of numbers – a view which he himself also held to be true. But he was equally aware that the mysteries for which Pythagoras was so venerated could not be summarized in easily communicated formulas or through vague rhetoric. If the Pythagoreans said that 3 is the key number to everything, because each thing is determined by a beginning, a middle and an end, one could not turn 3 into a perfect number – a number any more perfect than 2 or 4. The use of an epistemological formula as common currency, as a principle that could be applied on every occasion, would have made the secret of number a vulgar triviality – and truth a mere stereotype. Truth, instead, needed to be investigated by means of reasoning and mathematical demonstrations.

To be able to see in some epistemological formula a structure applicable to every domain of knowledge required a full and deep understanding of calculations, down to their smallest details – and the knowledge of how to penetrate beyond what Benedetto Croce would go on to mock, in the words of Giambattista Vico, as the study of the 'little geniuses'.[3]

And yet, when considered carefully, the kind of reasoning that inaugurated modern science continued to make use of intuitions recognizable from Platonic and Pythagorean formulas. Galileo himself did not exclude this eventuality. The study of falling bodies, on which the development of dynamics depends, led him to resort to a model that can be traced back to the very origins of mathematics and is intimately linked to the nature of the continuum: the existence of a case of equivalence between 'major' and 'minor' (*Dialogue Concerning the Two Chief World Systems*, I).

Analogous paradigms of thought, based on the principle of continuity, seem to have already been outlined in the ancient arithmetic of Mesopotamia,[4] and it is possible that they inspired the Euclidean concept of relation and proportion (*Elements*, V): the relation of commensurable or incommensurable quantities as entities made up of a series of smaller numerical relations plus a series of larger numerical relations. The methods for solving an equation numerically have always been based on a similar paradigm, and to emphasize this an algorithm well known since the remotest antiquity, the so-called *regula falsi*, or false-position method, was called by Arab mathematicians the 'rule of the pans of the scales'.[5] The sought-after unknown value operated as its axis, and within it was found the point of equilibrium of an indefinite number of oscillations of numerical approximations using the principle of more or less.

The pair of scales with unequal arms, the principle of the

lever by which commensurable magnitudes are balanced at distances in inverse proportion to their weight, was discussed by Archimedes – and Archimedes himself used the image of the balancing scale in the heuristic problem-solving expositions that preceded even the most rigorous demonstration. In the case of the squaring of a parabola, the heuristic method, which was mechanical in character, consisted of balancing the parts of a figure, which completed that of a segment of the parabola *P* and of an inscribed triangle *T* (with the same base and the same height), with the parts of another figure the measurements of which were known. It was demonstrated in the end that the area of *P* is 4/3 that of *T*.

The method of exhaustion used by Euclid and Archimedes was based on the more or less method that approximated an equivalence. To demonstrate that circles stand to one another as the squares constructed on their diameters (*Elements*, XII, 2), Euclid makes reference to two sequences of polygons, respectively inscribed within and circumscribing a circle the areas of which approximate those of the circle by excess and defect. The circle rests in the middle and corresponds to an ideal, yet impossible, equivalence between the inner and outer polygons. The possibility of getting incrementally closer to this ideal equivalence by increasing the number of sides of the polygon allows the thesis to be demonstrated.

The suggestiveness and efficacy of the image of the balancing scale appealed to the very divinities on Olympus, where Zeus used such a balance to decide fates during the Trojan War, and Athena to exercise justice during the trial of Orestes, both according to it a central decision-making role between contrasting outcomes.

Large-scale modern algorithms for solving a system of equations, or for calculating the minimum value of a function, are still based on the same idea that inspired mathematicians who lived in ancient civilizations such as that of Mesopotamia (around 1800–1700 BC) or Vedic India. A Babylonian tablet dating to 7289 BC analysed by Otto Neugebauer and Abraham Sachs has a square with its diagonals and some numbers which seem to indicate knowledge in Mesopotamia of accurate approximations by excess and defect of $\sqrt{2}$. This knowledge was subsequently inherited, Neugebauer speculated, by Indian mathematicians.

The law of continuity used by Galileo to study the motion of bodies, from which in the seventeenth century the modern revolution in scientific method arose, can be found already prefigured in Mesopotamian calculations and Vedic mathematics. Without the timely understanding of those calculations, there would have been no knowledge of the ancients, or indeed of modern science and the calculus of the last century – and perhaps not even the metaphysics that from ancient Pythagoreanism, through the Platonists, developed throughout successive centuries in the West.

The reason for all of this may be found in the simple operations that it is assumed made it possible for a scribe to approximate $\sqrt{2}$. These operations seem to be dictated, as Neugebauer and Sachs conjectured, by the idea of constructing an indefinite sequence of increasingly small intervals, one inside the other, and with the extremities defined by two rational numbers, each one of which contains $\sqrt{2}$. The irrational number $\sqrt{2}$ would then be configured, from the time of Mesopotamian calculus in antiquity, as a mathematical entity defined by the sequence of intervals that contain it. Similar strategies for approximating relationships between incommensurable quantities, through relations between whole numbers, would go on to be developed in Greece – and the definition of a real number, at the end of the nineteenth century, would rely on similar models of calculation to those used in Mesopotamia. Neugebauer seems to be perfectly aware of this remote derivation. We should note finally that the hypothetical algorithm that would have allowed the scribe to approximate $\sqrt{2}$ is a specific example of a general method, refined between the sixteenth and eighteenth centuries, for solving numerically an algebraic equation of any degree – a method which not only serves to approximate an irrational number but is also a sort of *clavis universalis*, or universal key, a fundamental strategy of mathematical analysis applicable throughout history.[6]

4. Growth and Decrease, Number and Nature

The geometry of Vedic altars would have required, with hindsight, a study of the relation that occurs between numbers and geometrical figures. The relation *straight line–number* was connected to incommensurable magnitudes and to the solution of algebraic equations, while the relation *curve–straight line* would have led to the methods of exhaustion, to infinitesimal calculation and the study of transcendental numbers such as e or π.[1] In both cases mathematics was obliged to confront the irreality of the infinite, which had the potential to subvert all of our contacts with the world.

A decisive circumstance ensued: it fell to the gods themselves to demand a description of how quantities vary, how they grow and how they decrease. Growth and decrease threw into doubt the very essence of things themselves, the *tò tí ên eînai*, according to the Aristotelian formula (*Metaphysics*, 983 a 27–8), the *quod quid erat esse*: 'the fact, for a thing, to continue to be that which it was'.[2] To all intents and purposes, the very foundation of metaphysics implicitly required an explanation of how any given entity, although subject to change, could yet preserve its essential characteristics. The entity needs to be capable of being recognized, and this recognition represents the presupposition necessary for its definition (*horismós, lógos*).

The question of growth was widespread in the cultures of antiquity and had cosmological aspects that could be traced back to the movements of the sun and the moon. For the Orphics, as Proclus relates in his *Commentary on the 'Republic' of*

Plato (II, 58, 10–11): 'the entire annual cycle is divided between growth and decrease'. And the astral bodies, especially the moon, governed an analogous oscillation between growth and decrease in life on Earth: the moon influences all sublunary beings and provokes in everything, thanks to its power, growth and decrease (*Commentary on the 'Timaeus' of Plato*, II, 87, 20–28, 12).

The notion of growth was included within the broader theme of change and loss of identity during the course of becoming. Writing in the fifth century BC, the Sicilian poet and dramatist Epicharmus speaks of this in one of the first testimonies devoted to the doctrine of Pythagoras. Iamblichus relates how the dramatist refrained from philosophizing openly, preferring instead to present Pythagoras' ideas by veiling them in the form of diversions, of theatrical representations (23 A 4 DK). Epicharmus was among the first to raise, within comedy, the question of how present in the phenomenon of growth is what the Greeks called *állo* – that which is always other, different, similar to formless and changing matter, an unstable and unreal flow. His characters, just as later in the case of the elderly Strepsiades in Aristophanes' *The Clouds*, are required to pay a debt incurred years previously – but deny being debtors because they are no longer the same people they had been before: an argument that smacks of sophistry. A fragment explains the issue more explicitly (23 B 2 DK):

> So now also consider mankind: as one in fact grows up, the other instead declines:
> All, in short, are changing, all of the time.
> Now, that which by nature must change, and never remain the same,

Is already different from what it was.
And you, in truth, and I as well, were yesterday other than we are now,
And will be different again tomorrow: never the same, according to the same law.

Mutability is a given, but it can happen according to a law (*lógos*). We can conjecture that this law has a mathematical meaning and that it depends precisely on the notion of relation, by means of which diverse things are interrelated – or are different parts of the same thing. We receive a first confirmation of this from Aristotle, who affirmed (in *Meteorology*, 379 b–380 a) that 'for all the time that a certain relation (*lógos*) lasts, the nature (*phýsis*) of a thing remains unchanged'. Starting with Pythagoras and with the *Śulvasūtra*, mathematics has always aspired to this duration or invariability during the process of change – just as, over the course of the centuries, the Platonic philosophy that took its inspiration from mathematics has.

Growth may be unlimited, and decrease may be as well. If there is no limit, with growth we move in the direction of the infinitely large, and with decrease towards the infinitely small – and in both cases what ultimately prevails is the non-being of the *ápeiron*, with the entity removing itself from our capacity to conceive of it, becoming indefinable. But it is not only the infinite that entails this kind of annihilation. In more recent times we have ascertained that a mathematical entity also verges on undefinability during the finite process of calculation, due to an abnormal and uncontrolled growth of numbers.

If we turn to Greece, a key concept regarding everything to do with the growth of quantities was that of *phýsis*. But it is misleading to translate *phýsis* as simply 'nature', without further qualification. Aristotle (in *Metaphysics*, 1014 b 16–1015 a 19; *Physics* 193 a 28–193 b 21) goes a good deal further, surveying various possible meanings, and specifies that nature must connect, above all, with the ideas of growth and generation. Nature, Aristotle asserts, is 'the first immanent element from which follows all that grows' (*Metaphysics*, 1014 b 17–18); and also 'a principle and a cause of motion and rest for the thing in which it immediately resides, for itself and not by accident' (*Physics*, 192 b 20).

The Aristotelian concept of nature appears to move along two distinct tracks: one which may refer back to prior materialistic theories, the other in keeping with an idea of motility governed by form (*morphé*) and *lógos*. Regarding the first of these, Aristotle reminds us that some philosophers contend that the nature and the reality (*ousía*)[3] of its products reside in their materiality (*Physics*, 193 a 10–12). The name Aristotle uses to denote this materiality is *prôton arrýthmiston*, the first constitutive element of a thing, still devoid of form, like the wood of a bed frame or the bronze of a statue. *Arrýthmiston* – badly proportioned, without rhythm – is the negation of *rythmós*, the term which unites the two notions of movement and form and which, before assuming the meaning 'rhythm', could mean in a more general sense the form adopted by that which is mobile, fluid and subject to modification (Chantraine, Benveniste).

Aristotle emphasizes, then, the meaning of *phýsis* as at the root of the provenance of things that are still devoid of form and incapable, by themselves, of undergoing a change taking them beyond their own virtuality or potential (*dýnamis*). Nature may be understood, he explains, in the sense of *hýle*, or the Latin *silva*, as being *matter* as opposed to form. But

Aristotle also opposes to this another point of view, according to which we can conceive an idea of nature based on form itself (*morphé*), and on the aspect it manifests according to the *lógos* (*tò eîdos tò katà tòn lógon*).

The comparison of these two positions represents a salient point in the history of our ideas about nature. Heidegger highlighted as much, commenting that the vision of nature as *hýle*, as matter, had the effect of divesting from movement every characteristic pertaining to invariability and permanence. If the matter of which things are made is water, fire and earth – and if the constitutive elements are only atoms – then the flux and the movement of *phýsis* fall under the sign of mere evanescence and inconsistency. In keeping with this vision, 'everything that has the character of movement, every alteration and every variable state [*rythmós*], ends up among what only occurs accidentally to the entity; being something which is unstable, movement is a non-entity'.[4]

So what, on the other hand, was nature like when seen instead precisely as movement and the growth of form, as *rhythmós*, as a figure or appearance conforming to the *lógos*? Heidegger himself pointed out that, in the language of mathematicians, *lógos* means something akin to relationship and rapport, that *lógos* is the substantive of *légein*, meaning above all to collect using a process of discrimination, and hence, we could add, to select single discrete operations in order to make only one, articulated and united with numbers – to bring through number 'a unity to scattered things'.[5] How does mathematics act with numbers and relations? Relation unites two quantities that are separate, making them one, as it were, in the process of looking for the measure that is common to both. In an analogous way, an equation correlates different quantities in the same formula; the mathematical concept of the set aims to gather different things into a single collection; the method of recursion collects together a variety of operations – for instance, all the possible additions between numbers – in a single concept of operation, the *sum*.

Hence Aristotelian discourse alludes indirectly to the phenomenon of metamorphosis, to the question of transforming identity in the process of change, and the possibility of escaping the accidental nature of becoming – by way of relatively stable configurations, where *lógos* and *morphé* ultimately prevail. Each thing becomes a wavering image, declares Pythagoras in Ovid's *Metamorphoses* (XV, 178): '*cuncta fluunt omnisque vagans formatur imago*'. But if we look at numbers and geometrical figures, it becomes clear how the process of change can be regulated by a relation, and how a quantity might be grown without altering its basic form. Only if the form remains stable is there the possibility of recognition in the process of growth: that which was before continues to be now. The Aristotelian thesis becomes consequently even more significant, whereby the permanence of a relation (*lógos*) guarantees the constancy of the nature (*phýsis*) of a particular thing.

A passage from Plato reveals just how much stability of form was to be considered a divine prerogative (*Republic*, 380 d):

> What can one say, then, of this other law [*nómos*]? Do you think perhaps that the god is a kind of magician who is capable, for the pleasure of deceiving us, of appearing before us at one time in a particular form and at another in a different one, or of changing his aspect [*eîdos*] into a crowd of different figures with the aim of deceiving us into thinking that he is like this?

That 'changing his aspect into a crowd of different figures' (*alláttonta tò autoû eîdos eis pollàs morphás*) renders the drama of metamorphosis and presents it as the inconstant and changeable antithesis of the face of the divine. In opposition to this, the god appears to be recognizable and *real*, as in the epiphanies of Artemis on the summits of mountains, of which Homer speaks in the *Odyssey* (VI, 102–9).[6] So the *eîdos*, the long-awaited vision, far from resembling the elusive and phantasmal idea of modern empiricists, does not consist of deception or magic but of exact and truthful experience – and therefore cannot simply disintegrate into a flux of incoherent and discordant images. Everything grows, diminishes and is transformed in the flow of *phýsis*, but the god shows his face to be unchanging. Perhaps it is for this crucial reason that relations of equivalence were frequently sought among different geometrical figures, and the altars of the gods – of Apollo in Greece, just as of Agni in India – had to be capable of being altered in scale while maintaining their form unaltered. Mathematics took on the task of making possible this invariability that was a prerequisite of a recognizably divine figure. Proteus, the Old Man of the Sea, who was subject to continual metamorphosis, was not an exception to the rule, because he alludes by contrast to a principle of veracity: he emerges here and there from the indistinct plurality of marine currents, a corollary of his mutability, with the same intensity and precariousness with which one may grasp in the sea of existence, in a brief flash, the immutability of the Platonic Good.[7] Homer assigned to Proteus, moreover, the knowledge of numbers and of numerical reckoning (*légein*) (*Odyssey*, IV, 411–13).

But growth and decrease were not only generic processes and could refer to phenomena of expansion and contraction generated by an initial nucleus, by an originary form that remains unchanged during the process. This circumstance made it possible to connect *phýsis* to mathematics. This also gave rise to a *technique* for magnification (or reduction) that would last for centuries – a technique that is recognizable even in the most advanced algebraic *computatio*.

In Greece the problem of enlarging geometrical figures crops up in a variety of anecdotes. From Eratosthenes' letter to King Ptolemy we learn that Minos, the king of Crete, needed to double the size of a royal tomb in the form of a cube, and that the construction at Delos of a cubic altar twice the size of the existing one was prescribed by the oracle of Apollo as a means of warding off a plague. Even the celebrated Colossus of Rhodes had to be doubled in size. In Greece there is no surviving treatise on the problem of enlarging geometrical figures, but to the anecdotal evidence afforded by the Greek tradition may be added a factor that is hardly accidental and which allows a conjecture about the origins of mathematics to emerge more clearly: the systematic combination of mathematics and religion in the ritual thought of Vedic India. The theme of scaling up altars in the *Śulvasūtra* allows us to decipher the meaning of what happened in Greece.

The significance of the duplication of the cube, Eratosthenes explained, was to be found in the idea of the growth of forms. It was a question of developing from an initial nucleus an indefinite progression of figures on an ever-larger scale. Growth always implies change *into something else*, a movement that alienates and displaces but one that can also be realized in sequences of similar figures. The technique of enlargement of a geometrical figure in effect allows us to construct, from a triangle or a square, a larger figure that is still a triangle or a square. We will be able, in the most general way, explains Eratosthenes:

> to change into a cube every given solid shape that is delimited by parallelograms, or alternatively make it pass from one form to another [*ex hetérou eis héteron*] and make it similar [to a given figure] and magnify it [*epaúxein*] while respecting similitude, and to do this when we are dealing with altars and temples as well . . . But my invention will also be found useful by those who wish to magnify [*epaúxein*] the dimensions of catapults and other ballistic weapons, in which everything must increase [*auxethênai*] in proportion, width, length . . . if we desire the firepower and trajectory to increase proportionately.[8]

For Plato, mathematics served above all to 'facilitate the radical conversion of the soul from the world of becoming to that of truth and being' (*Republic*, 525 c), but he also claimed that in numerical progression you could find the stamp of *phýsis* – of nature. Both typically consisted of a flow, of a progressive genesis of forms beginning from a 'seed' capable of expanding and reproducing itself without limit. The problem of doubling the size of a square, as plausibly proposed in Plato's *Meno*, had the following significance: not only to exemplify a process of learning by way of *anámnesis* but also to reproduce in ideal figures the force of *phýsis* in a potentially infinite growth of similar forms.

We need to connect to *phýsis* the Platonic conception of the geometrical point, which was understood, as Aristotle points out (*Metaphysics*, 992 a), as the starting point and the generating principle of the line. Plato did not accept the idea of a point as a static minimum, as a constituent part of a line or a volume. A. E. Taylor rightly sees in this the adumbration of a concept that was to emerge later, in Greek Platonism – of the line understood as the *fluxion* (*rhýsis*) of a point, a term which Newton would introduce to the English language to express the central idea of the calculus that was being developed in the seventeenth century.

Newton considered mathematics to be a method rather than an effective explanation of the order of things, and hence avoided any definition which sought to explain the nature of *instantaneous motion.* Nevertheless, he did not sidestep this concept, making it the foundation of his *Methodus fluxionum et serierum infinitarum*, or *The Method of Fluxions and Infinite Series*, written in 1671 and first published in 1736.[9] In this treatise Newton described the variable quantities not as aggregates of infinitesimal elements, but as entities generated by the continuous motion of points, lines and surfaces – and called 'fluxion' the instantaneous speed of generation. In the simplest case the term *fluxion* denoted the instantaneous speed of a point that moves along a line. Subsequently, after 1690, Newton attempted to give full rigour to the theory of fluxions, defining its principal concepts in terms of that analytic geometry the supremacy of which he championed in comparison to other disciplines. He then proposed two definitions that help to show even more clearly the importance of the ideas of growth and diminution for the general history of scientific thought in the West:

1. *Fluens est quod continua mutatione augetur vel diminuitur.*
   (A fluent is what is increased or diminished by continuous change.)
2. *Fluxio est celeritas mutationis illius.*
   (In fluxion is the swiftness of that change.)[10]

Hence *fluens* was that which increases or decreases by means of a continuous movement, and *fluxio* was the speed of that movement. The fundamental concepts of analysis, for Newton, had their origin in the ideas of growth and diminution related to the instant of their emergence. Only later, in the nineteenth century, did the completely different conception of the fundamental ideas of calculus impose itself, giving it not a dynamic but a static and atomistic character. The potential and dynamic infinity represented by the formalisms of differential calculus of the seventeenth century was replaced, thanks to a rigorous notion of *limit*, by an actual infinity (a 'completed' series consisting of an infinite number of members) that was more likely, it was hoped, to lend a *real* existence to the continuum.

For Plato, just as for the Pythagoreans before him, numbers made possible a kind of unifying activity of consciousness with regard to the multiplicity of perceptions. It is argued in the *Theaetetus* (184 d) that the multiplicity of particular perceptions resides within us in a confused way, without converging 'towards a single determined form, whether that of a soul or whatever else one might wish to call it, with which . . . we perceive what is perceptible'.

How does the soul operate? The being that is present in all things, says Plato in the *Theaetetus* (186 a–b), is precisely that towards which the soul itself inclines, establishing relations (*analogízein*) between antithetical things and juxtaposing in itself the past and the present with the future. Hence one of the functions of the soul lies in establishing the nexus between two distinct perceptions that come one after the other in time, and the ideal connection is then that which leads us to recognize the *same* thing in its transformation into something *other*. Only those things are intelligible which, although subject to change, remain recognizable to us. In the absence of a numerical law that allows the recognition of changeable entities, the flow of perceptions degenerates into irremediable chaos.[11]

Now, with mathematics, it is possible to think of capturing existence in a stable link between distinct perceptions. Among twentieth-century mathematicians, Brouwer understood perfectly that the first phenomenon from which thought assumes its form is a temporal sequence in which our consciousness retains a primary sensation in the act of moving to a subsequent one. For Brouwer the phenomenon is repeated recursively, and our consciousness, distinguishing the past sensation from the present and standing back from both, gives rise to the activity of our mind. A mathematical form of this process can be found precisely in the procedures of iterative calculation, in which the variable approaches the solution of a problem with a succession of approximate values, each of which is calculated from the previous one by the repeated application of an operator. Incidentally, Norbert Wiener thought of analogous iterative mechanisms in the simulation of the behaviour of living organisms, especially in cases in which the phenomenon of feedback occurred. The grasping of an object, he noted, consists in the progressive reduction of the residual distance between hand and object, exactly in the way that in iterative processes, with their repetition of a series of operations, the residual distance between the value of the variable and the solution sought is gradually reduced by degrees.

5. *Katà gnómonos phýsin*: The Nature of the Gnomon

So how should we think about numbers, in order to accomplish the task of linking together the soul's perceptions? Numbers can be thought about in a variety of ways, but there is a fragment by the pre-Socratic Philolaus that shows us the right way – the most appropriate, that is, to our capacity for aggregating our thought, and the one from which the iterative algorithms of mathematics originate. Numbers, according to this fragment, 'harmonizing all things with the internal perception of the soul, makes them recognizable and commensurable among themselves, according to the nature of the gnomon [*katà gnómonos phýsin*], because it forms and disassembles all the individual relations between things, those that are without limit as much as those that are limited' (44 B 11 DK). Here we find all the decisive elements: numbers, the soul, things (*prágmata*), perception, knowability, relations, the notion of the finite, the infinite and, finally, the nature of the gnomon. The gnomon, Heron of Alexandria would go on to explain (*Definitions*, 58), is the figure which, when applied to a given geometrical shape, generates another shape similar to that first shape. Pythagorean numbers were arranged in a certain *geometrical order in space*, and their growth was in keeping with the nature of the gnomon: the number 4 consisted of four points arranged in a square; surrounding them with five points arranged along two of its sides, one obtains the subsequent square number, that is to say, 9. In the same way one obtains

16, 25, 36, 49 . . . adding each time an odd number of points arranged to form a square. Thanks to George Thibaut's translation in the 1870s of Baudhāyana's *Śulvasūtra*, we can perfectly understand the analogy between the bricks of the Vedic altars and Pythagorean numbers. The instructions for building the altar of Agni consisted of the following: start with a small square made up of four bricks; then, by adding five bricks, go on to make a square of nine; then again, with the addition of seven bricks, follow it with a square of sixteen.[1] Whoever established the measurements of the altar evidently conceived of the potentially infinite growth of square numbers.

Why does this way of arranging numbers fulfil the pre-established purpose of making things knowable? In order to

Figure 2

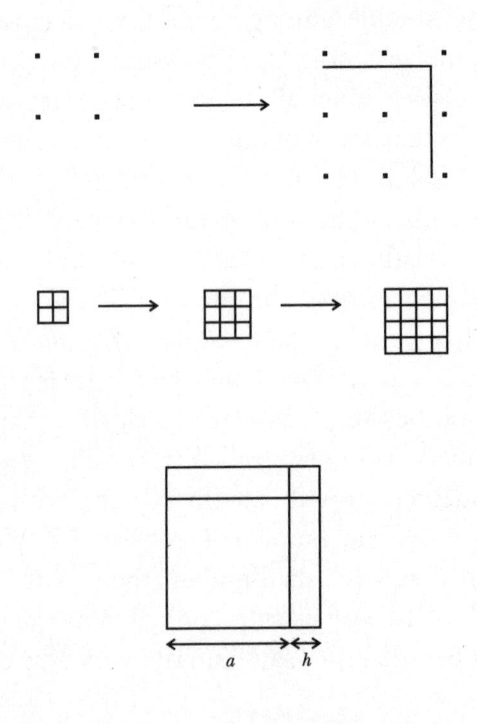

understand, it is necessary to reproduce in the geometrical continuum the idea of enlargement through gnomons, thinking in particular of the square. The image of just such an enlargement was used by Euclid to demonstrate the *geometric* version of the *algebraic* formula for the square of a binomial: dividing a straight line into two segments, a and b, the square on the whole line $a + b$ is equal to the square of a plus the square of b, plus two times the rectangle of the sides a and b, a theorem that is expressed by the simple formula of enlargement of the square proposed in Āpastamba's *Śulvasūtra*: $(a + b)^2 = a^2 + 2ab + b^2$. Newton and Raphson would go on to use this formula to resolve numerically a quadratic equation such as $x^2 - 2 = 0$. Given/supposing that $x = a + b$, where a is an initial approximation of $\sqrt{2}$, and disregarding b^2, the equation becomes $a^2 + 2ab = 2$, from which it follows that $b = -\frac{a^2-2}{2a}$. And from this we get the iterative method $x_{k+1} = x_k - \frac{(x_k^2 - 2)}{2x_k}$ where $k = 0, 1, 2 \ldots$ and $x_0 = a$.

Euclid's theorem (*Elements*, II, 4) implies a great deal more than it actually states, and we may wonder how far Euclid himself was actually aware of this. The sense in Philolaus' fragment of the expression *katà gnómonos phýsin* needs to be connected with the fundamental meaning of *phýsis* as involving *growth*, and indeed Euclid tells us how a square grows when a side a is added to by a segment b. Interpreted in this way, the meaning of the theorem is radically altered. The segments a and b do not have the same meaning, because b is interpreted as an *increment* of a. From here the analysis begins: in general, we ask how the value of a function $f(x)$ changes if x undergoes the incremental change b. Euclid (II, 4) shows us the case of the simple function $f(x) = x^2$, and we ask ourselves how the square of x varies with the variation of x. But the problem is altogether general in scope: it is about how a growth (or a diminution) occurs to a quantity A that depends on another quantity B when the latter undergoes an increment, positive or negative. The first comprehensive answer to the problem is to be found in the celebrated treatise *Methodus incrementorum directa et inversa* ('Direct and Indirect Methods of Incrementation', published in London in 1715), in which Brook Taylor introduced a formula which expresses how a function $f(x)$ varies for small increments of the variable x. Lagrange would later define Taylor's formula as the fundamental principle of differential calculus. In one direction or another, the increments of x can determine the decrease of the function f, and we thereby obtain a class of methods for finding the value of x by which f assumes its minimum value. We can gauge the importance of this from the celebrated words of Leonhard Euler, according to whom the perfection of the universe was expressed in the fact that nothing happens in the world without some rule of minimum or maximum being involved.

It is difficult to imagine how it would be possible, in the absence of such an analytical device, to measure the effect of the disturbance that a quantity undergoes following the alteration of another quantity. In the analytic formulas we find the first instruments for studying the stability of a system of calculation faced with disruptive errors in the data and in individual operations. The formula for the square of the binomial makes it possible to consider two successive moments of change, from a to $a + b$, and to rediscover another square from an initial one. This implies a stability, an invariance in change on which the *epistéme* is based, the power of the mind to dwell on something. Change is not only *in(to) something else* (*eis állo*);

becoming is not just non-being. In this view, *phýsis* becomes a flow of iterative generation of similar forms, or at the very least we are convinced that this might be so – and examples are studied in which it is indeed the case. But already at this juncture an algebraic calculus announces itself, modelled on the growth of geometrical figures, in which the phenomena of the growth and generation of *phýsis* might have found a mathematical representation. And something like the same rule applies to our everyday experience: without repetition, as Proust has it, unreality prevails.[2]

Extrapolated from the formula for the square of the binomial in order to solve an algebraic equation numerically, the methods developed by Viète, Newton and Raphson are justly celebrated. This was the most powerful and efficient machine for solving equations and calculating the minimum value of a function – something mathematics found itself re-elaborating, in modern times, following computational laws laid down in antiquity. Just a few decades ago it was demonstrated that Newton's method for approximating the square root of a number is the most efficient possible, measured in terms of speed of convergence and computational complexity. Equally recent is the development of methods of calculation of the minimum value of a function which generalizes Newton's schema.[3]

At this point, however, we should not neglect to mention a crucial characteristic, which is manifested in digital calculation and in many other forms, and is one of the main causes of the instability of numerical algorithms. In the fractions p/q generated by Newton's method for approximating the root of an equation, the numerator p and the denominator q can rapidly grow beyond tolerable limits with a consequent increase in the computational cost, and with a loss of information due to the process of rounding off.[4] In the calculation of square or cubic roots the growth of numbers is linked to the growth or diminution of a geometrical shape through successive gnomonic corrections, and, in the original logic of the process, to a progressive diminishing of the unit of measurement.

Today the strategies of numerical optimization for calculating minimum and maximum values of functions are based on the same scheme of methods used by Viète and by Newton. And on these same strategies depend the processes of automatic learning with neural networks, which are based on the calculation of the parameters for progressively reducing the gap between calculated and expected answers. In this sense, learning, or better, the model of learning realized by the neural network, is conceptually identical to a process of minimization. But analytical calculation is not always able to guarantee the success of the operation. In the automatic processes of learning, uncontrolled growths of numerical values may often occur, due to the ill-conditioning of matrices, with a consequently fatal loss of meaning for the results of the calculation.

6. *Dýnamis:* The Capacity to Produce

In ancient Greece, through the enlargement of a square and other geometrical figures, an active production was expressed, almost a germination, such as the one to which the *phýsis* of the Orphics alludes. An analogous idea is found in the *dýnamis* (potentiality) by which a surface extends outside a line. Euclid showed how to construct a square on a straight line, in the sense of extracting a figure *from* one side.[1] Proclus (*Commentary on the First Book of Euclid's 'Elements'*, 423–4) would later remark on the difference between constructing (*systésasthai*) a triangle and tracing (*anagráphein apó*) a square by almost *producing* it from a single side. In Mesopotamian mathematics, a term of uncertain meaning – *takīltu* – occurs that could well mean, according to Thureau-Dangin's hypothesis, 'that from which something has been produced', like a number that when multiplied by itself produces a square.[2] Pythagoras' theorem, the demonstration of which is to be found in Euclid's *Elements* (I, 47), could well be a response to the criterion of the growth of geometric figures that keep their form unaltered. From two squares we build a larger third, the sum of the first two, producing it from the diagonal of the rectangle the two sides of which coincide with those of the two squares. Growth and production seem to be united by the same law.

Pythagoras' theorem was known in ancient Mesopotamian mathematics, although there is no record of a demonstration of it such as Euclid's. The Babylonians tried to solve, by means of algorithms, simple concrete problems that did not possess the degree of abstraction and generality that we encounter in Euclid's theorems. Instead the Babylonian calculations show a command of computational schemas that are as efficient as they are enduring, and which are still today the basis of the most advanced calculations we use to solve complex mathematical problems. A problem typical of Babylonian arithmetic consists of calculating the diagonal of a rectangular door of which the height and width are known. The scribe connects the solution back to the problem of the enlargement of a square, which suggests a scheme of calculation analogous to that which modern mathematics would adopt in order to solve an algebraic equation.[3] Consequently, in order to observe the criteria of calculation, Pythagoras' theorem is explicitly connected to the growth of a geometric figure.

In Āpastamba's *Śulvasūtra* (I, 4–5) the same theorem of Pythagoras is formulated in such a way as to show that the diagonal of a rectangle produces (*karoti*) outside itself that which the smaller side and the larger side each produce (*kurutas*) externally.[4]

In the Greek term *poieîn* (to produce) we catch the elementary principle of *building*, analogous to the Sanskrit *vi-hṛ*[5] geometrical figures of increasing dimensions, according to criteria that have marked ancient geometry as well as numerical calculation in both antiquity and the modern era. In *poieîn* there is the idea of operating efficiently: we can for instance calculate in an effective way the expression of a multiplication, or of what we call today – not accidentally – the 'product'.

In Greek mathematics the term *dýnamis*, which refers to the capacity to be or to become something, denotes the side of a square – that is to say, the square root of the number that measures the area. Hence the lines connoted

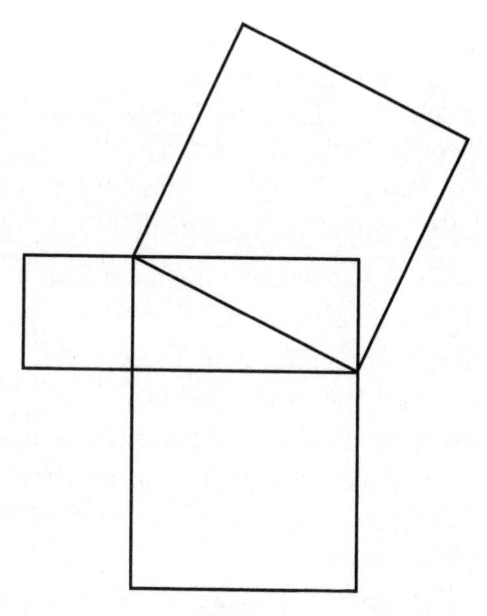

Figure 3

by a *dýnamis* may also mean the squares for which they compose a side: the square root of 3 can denote the square with an area of 3 (*Theaetetus*, 147 d ff.).

In the *Statesman* (266 a–b) Plato mentions the diagonal, and the diagonal of the diagonal, as if it was a question of indefinite progressive generation, of the doubling of a square that the initial side has the *power* to generate.

In the Orphic and Pythagorean traditions (Philolaus, 44 B 11 DK and 315 Kern) this idea of power was also contained within that of the perfection of the decad, which was the 'principle of divine, celestial and human life', without which 'all things would be limitless, obscure and uncertain'.

The notion corresponding to *dýnamis* in the mathematics of Babylonian antiquity was *mithartum*, the geometric square produced by its square root (*íbsi*).[6] In the geometry of the

Vedic altars the generation of successive squares, one from the diagonal of the other, is rendered by the term *dvikaranī*, the diagonal which produces the double,[7] or, according to the indication of Plato's *Meno*, an indefinite succession of squares, one after another, that is to say, with the diagonal of one equal to the side of the next. In all of these cases the phenomenon of *production* is essential: a mathematical entity *produces* another, as if what was entailed consisted of a power of progressive enlargement inherent in those entities themselves. Similarly, at a much later date it will be claimed that the arithmetic of fractions produces phenomena that allow us to recognize and to define new numerical entities: real numbers, defined by Dedekind as *sections*, will be *produced* by properties of only rational numbers.

Heidegger grasped the meaning of *poiésis*, of production, as the central nodal point of Greek *téchne*, specifying that every production is based in this context on the truth of 'unveiling'. Quite different, according to Heidegger, is the case of modern 'technique', in which the unveiling is not a production (*poiésis*) but a provocation (*Herausfordern*), which results inevitably in the attempt to exploit energy on

Figure 4

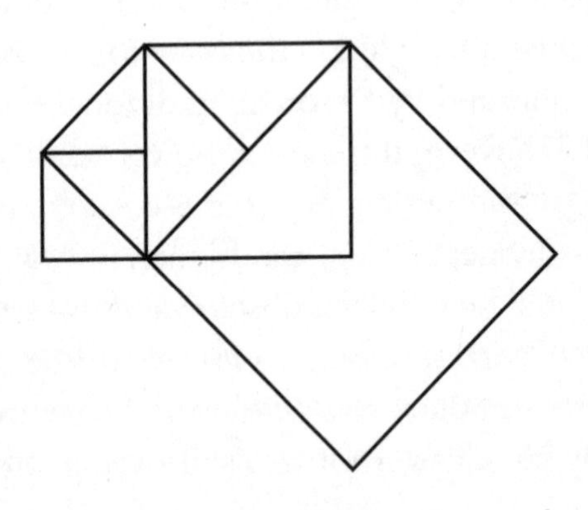

an industrial scale, according to the degenerate criterion of maximum utility with minimum cost.[8] The algorithms that allow this critical transition, this degeneration of the process of production, still rely upon the same elementary operations of ancient mathematics, the ones delegated, precisely, to production. The revitalization of the meaning that these operations possessed in the knowledge of antiquity resulted in the extraction, so to speak, of their most arid aspect: the one most instrumental for the aims of technological functionality.

The ontological implications of mathematical *dýnamis* emerge clearly from a passage in Plato's *Sophist* (247 d–e):

> I therefore assert that everything which by its nature possesses a capacity (*dýnamin*) to produce (*poieîn*) any effect or to be subject to it, even the most insignificant thing and to the smallest degree, and even if only once, all this exists in reality (*toûto óntos eînai*). I propose in fact a definition: entities are nothing other than power (*dýnamis*).

The course of being, says the Eleatic Stranger in the *Sophist*, entails precisely an act of production, and that which is brought into being must be defined as a 'product' (*Sophist*, 219 b). Socrates in the *Republic* (477 c) further noted that in *dýnamis* we can find the type of things (*génos ti tôn ónton*), an observation that was repeated verbatim by Proclus (*Commentary on the 'Republic' of Plato*, I, 266, 17). The Eleatic Stranger of the *Sophist* (238 a) placed 'the entire province of numbers among the things that have being', and the meaning of *dýnamis* as a power of extension and as a principle of natural generation, as the moving agent of a *phýsis* which literally germinates entities on the model of numerical progression, is made explicit in Plato's *Epinomis* (990 c):

> Now the primary and most important study is of numbers in themselves, not of corporeal numbers, but of every possible generation and potentiality (*dynámeos*) of growth [with elevations to the square and to the cube], of the odd and the even, and of the influence that numbers have on the nature (*phýsis*) of things.

In the *Philebus* (15 d; 25 d–e) Plato had seen a paradoxical possibility of *coincidence* of the one and the many by way of the *lógos*, that is to say, of mathematical relation (*tautòn hèn kaì pollà hypò lógon*), and had ascribed to numbers and relations the capacity to end the oppositions between contraries. In numbers, therefore, what is expressed is not a mere multiplicity but on the contrary the realization, thanks to relationality, of the synthesis between the one and the many that is inherent in *actual* things. This, Plato declares, is knowledge bequeathed to us by the 'ancients, who were better than us and lived closer to the gods' (*Philebus*, 16 c). In the *Timaeus* (52 c) he adds, furthermore, that it is the *lógos* that is responsible for telling us with truth and exactitude 'that which it really is, showing that, as long as a thing is a thing and another thing is another thing, then neither of the two could enter into the other so as to remain itself and at the same time become two'. Hence the *lógos* is not simply a discourse, but in a more specific way the *relationship* or interplay whereby mathematics is given the role of constructing appropriate definitions. The mathematical relation is the instrument which interrelates one thing with another, putting one and two together, exactly as is demonstrated by a progression of similar figures which are many and yet, due to the invariability of the form, one and the same. Similar progressions can typically intervene in the same definition of relation, according to Aristotle's indications (*Topics*, 158 b), as in the case of *anatanaíresis*, or in the Euclidean algorithm for the calculation of the ratio of two magnitudes. Without a knowledge of numbers, Plato believed, we would lose the possibility of knowing the real and be reduced to only perceptions and memories (*Epinomis*, 977 c).

Against potential pulverization into the manifold, numbers were the last defence of an unfolding existence. It is not surprising, then, to find that in the fragment by Archytas, it is claimed that the science of calculation has, with respect to knowledge, a clear-cut superiority in comparison to other disciplines and is capable of tackling anything it wants in a way that is clearer and more obvious (*enargéstero*) than geometry itself. The Greek term *enargés* refers to something that appears and is graspable in a transparent, palpable way, almost akin to a corporeal form. In exceptional cases the gods themselves could adhere to such apparitions (*Odyssey*, XVI, 161). Iamblichus would go on to observe that Pythagoras had defined number as the *extension* and *actuation* of the seminal *lógos* immanent in the unit.[9] An important testimony that helps us to understand the significance of numbers as a process of growth inherent in *phýsis*; a process that takes place gradually, the Orphics were wont to say, like the meshing of a net (Aristotle, *On the Generation of Animals*, 734 a) or in a sort of progressive ramification (316 Kern).

Taken together, these clues seem to suggest that numbers were conceived of as a physical entity, though certainly not for lack of a capacity for abstraction. On the contrary, number was the *res vera*, the ultimate reason behind the actuality and reality of things, the principle of the evidence of their existence in the world, against all possible arguments in favour of the illusory nature of the real. For this reason, the function of the link between the human and the divine belonged to numbers as entities capable of representing a command of evident, intangible truth.[10]

As Simone Weil wrote: 'consciousness and reality are proportional to the multiplicity of systems captured simultaneously in a single operation of the spirit'.[11] Now it has always been numbers that have allowed this operation to take place in a variety of ways, regardless of how the notion of number was understood in different contexts over the course of centuries: as a principle of combining different entities of the same class; as the synthesis of varied, individual operations in a single function definable recursively; as a real number, atomic, an element in a dense abstract field, ordered and complete; as a section of the rational field, analogous to the Euclidean concept of relation; as a number conceived according to the nature of the gnomon, which fulfils a nexus between two distinct moments in the growth of a geometrical figure, thereby allowing the same figure to be recognized after undergoing the changes to which it has been subjected. And finally, there is the notion of number as a computational process, as an algorithm that must respond to precise requirements of efficiency.

Plotinus asserted that all intelligible things are in a certain sense also real things (*kaì noerà kaí pos tò prâgma*). So it was for surfaces, solids and all the shapes, in regard to their *where* and their *how*:

> Shapes, in fact, are not only conceived by us. They are attested to by the shape of the universe that comes before us, no less than by the other natural shapes [*physikà schémata*] that are in the things of nature [*phýsis*] that exist by necessity [*anánke*] above, before bodies, in their purity, as primary forms [*prôta schémata*]. (*Enneads*, VI, 6, 17, 20–25)

The decisive factor, the principal reason for the bond that united mathematics and reality, was that 'the infinite is in contrast to number' (*Enneads*, VI, 6, 17, 4) (even if we say that number is infinite, because every time we think of one we can also think of a bigger one). The notion of number and the relation between numbers were characterized by a limit, and they were consequently the opposite of the *ápeiron*, of that which is without limit, synonymous with absence (*stér-esis*), unreality and non-being. For Plotinus, there were two forms of infinity: one was the model, the other the image, and it was the latter that could most properly be called infinite (*Enneads* II, 4, 15, 20–25). Matter (*hýle*) was infinite because of its opposition to *lógos*, that is to say, to relation or connection, which in the measuring of one thing with another rendered possible instead an ordered disposition of real things in space and time. It is from the concept of relation that the modern notion of number has been derived.

For the Stoics (*Stoicorum Veterum Fragmenta* [*SVF*], II, p. 328), nature was both the force that holds the constituent parts of the cosmos together and the generating principle of earthly beings. The principle was contained within the *spermatikòs lógos*, the seminal *lógos* on which everything depended, animate and inanimate, and which had its counterpart in the unity from which the sequences of relations that approximate irrational numbers emanate.[12]

For the Neo-Platonists as well it was commonplace to think of numbers in strict correlation to *phýsis*, in accordance with Platonic observations on how nature conformed to numerical progression (*Epinomis*, 990 e). The concept of *physikòs arithmós* is found in Plotinus (*Enneads*, VI, 6, 16, 45–6) and in Nicomachus of Gerasa (first–second century AD),[13] for whom nature, *phýsis*, was a fabric made up of numbers and relations involved in generative flows organized according to precise arithmetical laws. In the eleventh century the Byzantine monk Michael Psellus would come up with the idea of physical number (*physikòs lógos*) as a complementary concept to mathematical number. The former pertains more closely to living bodies, plants and animals because each one of these is born, grows and dies in determined temporal cycles.[14]

All of this grandiose complex of doctrines, of systematic allusions to the affinity between numbers and *phýsis*, was capable of being founded, presumably, on the fact that numbers need to be thought 'according to the nature [*phýsin*] of the gnomon': a factor that at the time, and even more so in the future, would render operative an abstract model of numbers and a system of algorithms devised for the mathematical description of nature. Over the course of centuries, the 'reality of numbers' would have answered not only to the philosophical vision of a *noetós* cosmos, of a divine and intelligible universe, but also to the real deciphering of nature through symbolic models. And for a long while after that the gnomon did not go on to reveal all of its potential efficacy. After Viète and Newton, however, the same paradigms on which the ancient Greek, Indian and Babylonian algorithms had been based served to define the structure of more advanced methods of the science of calculation. In order to penetrate into the deep interrelation between calculation and knowledge in the ancient world, philosophical formulas are not enough: we must also follow every computational detail to capture its conformity with and theoretical contribution to those very same formulas.

7. Intermission: Spiritual Mechanics

Two enigmatic and laconic statements on the nature of the soul have come down to us from antiquity. The first is attributed to Heraclitus: 'to the soul belongs a *lógos* that increases itself' (22 B 115 DK); the second to Xenocrates, who was entrusted with the direction of the Platonic Academy after Speusippus: 'the soul is a self-moving number'. Aristotle (*On the Soul*, 408 b 34 ff.) consigned Xenocrates' judgement to the category of theories that were completely irrational, but he did not consider it superfluous to excogitate at length on his reasons for refuting it. To this may be added the intimate connection between a philosophy of the soul and a comprehensive theory of the nature of the universe that is prefigured in a question raised by Socrates (*Phaedrus*, 270 c): 'But do you think it possible to know the nature [*phýsin*] of the soul to any sufficient degree if we dismiss nature [*phýseos*] as a whole?' The answer, self-evidently negative, indirectly correlates numbers and the soul, because Plato was wont to affirm that in numerical progressions one could read the *phýsis*, and the 'nature' of the soul was precisely its *phýsis* – in other words, to discern its growth in the *lógos*.

Heraclitus' words are untranslatable and lend themselves to various interpretations, depending largely on which meaning, from among its many meanings, is given to the word *lógos*. But in the apparently absurd sentence penned by Xenocrates, what for Heraclitus comprised the *lógos* can be, in more circumscribed terms, number. In both cases, be it the *lógos* of

Heraclitus or Xenocrates' number, neither stands still: the first augments; the second moves. Now *kínesis* (movement) is a word that for Aristotle refers to any kind of process during the course of which a thing may change aspect, nature or position. It is not inappropriate to contend that *kínesis* might also refer to numeration, to a measured movement, to a passing of time registered by numbers. And we know that among the earliest meanings of *lógos* there is the one that refers to selecting and surveying by way of enumeration.

For the Stoics the passions of the soul were regularly commensurate with nature (*phýsis*). They were thought to follow or, more frequently, to infringe on the *lógos*, that is to say, numerical measure. 'Against nature' meant, for Chrysippus, 'against the straight *lógos* according to nature [*katà phýsin lógon*]' (*SVF*, III, p. 94). For Zeno of Citium (*SVF*, I, p. 51) passion was an expanding or contracting of the soul beyond measure (*álogos*), as if the mathematical laws on which natural growth depended, precisely the natural growth that Plato saw effectively imprinted in numerical progressions, had been violated. To grasp the reciprocal affinity of number, soul and *phýsis*, if we leave aside the change in meaning that the term *nature* acquired during the course of centuries, and especially in the seventeenth century, we could resort to Spinoza's thesis that 'nature is always the same, and its virtue and power to act is everywhere one and the same' (*Ethics*, III, Preface).

Another peculiarity brings together the sentences of Heraclitus and Xenocrates: number moves by itself; the *lógos* augments itself. Movement and growth occur due to the intrinsic predisposition of the soul to grow and to change its own state independently, by means of successive expansions, just as in the development of a seed. Bruno Snell[1] noted that this possibility of autonomous development of the spirit was something still quite alien to Homer, who always attributes growth in physical and spiritual force to the intervention of a deity. Equally alien in Homer's eyes would be the notion of *prôton kinoûn*, the primary cause originating in the soul itself, as Aristotle would conceive of it.

But from what is this 'from itself' of the self-moving soul generated? The seed grows invisibly at the start, almost mechanically, by dint of a virtual force the origin of which has not been disclosed to us. Simone Weil speaks of a 'spiritual mechanics, the laws of which, though different, are as rigorous as those pertaining to mechanics itself'.[2] There is a reference here to Mark (4, 26–32), where the Kingdom growing automatically (*mekýnetai*, *automáte*) and unknown to us is compared to a seed germinating in the earth. But the soul is also tied to the body, as Plato pointed out (*Phaedon*, 83 d–e), and every pleasure and pain fixes and almost nails down the soul to a corporeal form, inducing it to believe that everything must be true that the body claims is true. If it were to follow this inclination, it could not then arrive pure in Hades but would fall into another body and grow according to its nature: 'as if it were a seed, it will germinate there [*hósper speiroméne emphýesthai*], and because of this will never participate in that which is divine, pure, uniform'.

This puts me in mind of the words of Epicharmus on how growth occurs subject to a law – words which in turn have a remarkable affinity with one of Goethe's lyrics:

> Just as on the day it brought you into the world
> the sun was up above greeting the planets,
> soon you grew bigger
> according to the law that regulated your beginning.

> And so you must be, you cannot escape yourself –
> so said the Prophets and Sibyls of old;
> neither time nor strength can break
> the imprinted form that develops throughout life.[3]

When they are generated by an iterative formula – from the simplest, which consists of adding a unit to a natural integer every time to find the next one, to the most complex recursive procedures – numbers, just as in the case of the soul, are subject to an automatic movement. Dedekind and Turing would both grasp the automatic nature of arithmetical processes, the first through the mechanism of recursion, the second with a more suggestive and explicit notion of the mechanical. Nevertheless, if not Dedekind then at least Turing was aware of the insidious possibility of an uncontrollable growth of numbers. He understood, in effect, how an error in the arithmetic of the machine can grow if the numbers being calculated become very big, and he quantified the growth in error, in the case of a system of linear equations, by mean of functions capable of measuring the *magnitude* of certain matrices.

If nature, as Nietzsche noted, is really excessive, 'wasteful beyond measure, indifferent beyond measure, without purpose and scruple',[4] if, in truth, this excessive character belongs to *phýsis* – an overwhelming, excessive influx of presence that Heidegger also attributed to it – then it is conceivable that numbers, which resemble *phýsis* in other respects, may also grow beyond all control. There was not just infinity on the horizon with the growth of numbers: with large-scale automatic calculus our gaze had to learn to stop well before it, on the firm and still little-known ground of the finite, be it very big or extremely small.

8. Zeno's Paradoxes: The Explanation of Movement

The most paradoxical revelation of the unknowable nature of the infinite originates with Zeno of Elea, in the form of dialectical passages that throughout the centuries have never really been refuted. The history of Zeno's paradoxes is made up of contradictory commentaries, critiques and refutations; of revisitations and surprising vindications that have ended up turning their meaning upside down: Zeno was right, the world needs to be rethought according to his arguments, and if we elaborate a mathematical theory of the continuum which is coherent and appropriate, the paradoxes reveal themselves to be ideal instruments for reinterpreting the real world. They are not idle speculation, they are reality, and they show the way in which mathematics can represent it. This, at least, is the explanation that seemed to prevail in the early twentieth century, with the development of a mathematical theory of the continuum, the origins of which date back to the second half of the nineteenth century.

The paradoxes on the subject of movement are based on the progressive diminution of the amounts of distance covered, and on the observation that the decrease *ad infinitum* of the segments in which a continuous line can be divided causes the world to appear to us as something unreal. It seems that the infinite (*ápeiron*) entered into Eleatic thought as far back as the sixth and fifth centuries BC as a dilemma because it is expressed in a double version: one affirmative, above all in Melissus of Samos; the other negative and apparently in opposition to the first, as in Zeno's paradoxes. Melissus expresses himself differently from Parmenides. For the latter the infinite is not an attribute of being, because in his view it is negatively marked by a *non*-being, the absence (*stéresis*) implicit in the Greek *ápeiron.* While for Parmenides the force of necessity seems to constrain being within the fetters of limitation (28 B 8, 26–33 DK), Melissus resolutely affirms that the being that has always been and will always be, without beginning or end, is infinite – and that nothing that has a beginning and an end can be called infinite (30 B 1–4 DK). The infinite, Melissus was wont to assert, is also *one* (30 B 5 DK), because if it were *two* it would have a limit in another thing (*pròs állo*).

The infinite, from Zeno's point of view, is an engine for generating paradoxes, with very different results: denying the possibility of movement, or showing the insufficiency of our rational explanations of what we perceive with our senses.

The dichotomous division we find in Achilles' chase has no end; it is the perfect example of a kind of potential infinity. If Achilles has to cross a space one metre in length, he must pass through the intervals of decreasing length $1/2$, $1/4$, $1/8$ . . . $1/2^n$, where *n* assumes infinite values. If the tortoise sets off with even the most minimal advantage, in the time it will take Achilles to cover that ground the tortoise will make a small step forwards; and if the pursuit is continued according to the same logic, then Achilles will never be able to catch up with the tortoise. What's more, Achilles is not logically able to cover even a metre, because he must first cover half of it, then half of what remains, that is to say a quarter of the whole – and so on, *ad infinitum.* If one considers for every point in the trajectory the infinity of intervals that need to be covered before reaching that point, one is obliged to conclude that Achilles, in fact, does not move at all.

If one looks at the laws of motion and at the simple formula that links the space covered s to the time taken t and to the speed v – that is to say, $s = tv$ – it becomes apparent that, if Achilles runs with a constant speed v, then he will overtake the tortoise that is moving with a constant speed v' that is less than v, even if the tortoise has an initial advantage s_0. In effect, there exists a value of t whereby tv, the distance covered by Achilles, is superior to $tv' + s_0$, that is to say, the distance covered by the tortoise.

A scientific basis for the obvious advantage that Achilles will gain over the tortoise also stems from the fact that the sum s of the infinite intervals into which the trajectory is divided is *finite*, being precisely equal to 1. If Achilles maintains a speed superior to a certain finite value v, he will complete the metre-long course in a time not superior to $s/v = 1/v$. He would not be able to complete it, however, if his speed diminished progressively and tended towards 0 like the length of the intervals into which the space is divided. But the paradox remains: how does Achilles manage to cover the infinite intervals into which a trajectory of a metre is 'segmented' in a finite time? Or, more generally, how can one execute in a limited time an unlimited number of operations? The argument based on the fact that the sum of the partial intervals is finite provides an answer to the question *when* will Achilles overtake the tortoise? But what still needs to be clarified is *how* he will manage to do this. How will he cross an infinite succession of intervals.[1] It still seems to be the case, and equally incomprehensible, that in order to cover a straight line Achilles must touch *all* of its points.

In referring to the paradox, Aristotle would reflect on the time it took to cover the distance (*Physics*, 233 a 21–31; 239 b 5–240 a 18). In the absence of a theory of limits pertaining to numerical series, he argued that, above all, a continuous line can be considered infinite in two ways: in relation to *division* or in relation to *the extremities.*[2] Only in the second case is the time taken infinite, while in the former, if the speed is sustained above a finite threshold, it is actually limited. His argument is sound, but it still lacks the conclusion that follows from observing that the sum of the infinite intervals of the course run is finite, and equal to 1.

Zeno develops his third argument against the existence of motion, relating to the flight of an arrow, in the following terms: if we accept that each thing is in a state of stillness or movement when it occupies a space equal to itself, and that the object only moves in the instant, then the arrow remains motionless.[3] Aristotle (*Physics*, 239 b 5–9) summarized the argument as if Zeno thought of an atomistic fabric of space and time, a continuum composed of infinite indivisible elements: 'The fallacy of Zeno's argumentation is obvious: as he alleges that a thing is still when it has not moved in any way from the space occupied by its volume, and in every fixed instant of the supposed course of its overall movement it remains in the space that it occupies during that instant, then the arrow does not move at any moment during the course of its flight. But this is a false conclusion, because time is not composed of actual instants, any more than every other quantity is composed of atomic elements.'

Hence Aristotle maintains that the arrow in flight remains stationary only if we suppose that time is made up of instants. It is in fact impossible for the arrow to move *in the indivisible instant,* because in that instant, by moving, it would change position, which would mean that the instant is really divisible.[4] But it remains unclear as to whether Zeno had assumed the atomistic hypothesis as a necessary given for the logic of his paradox. Even if one assumes that space and time are not composed, respectively, of indivisible points and instants, and that the infinite by division was entirely hypothetical, movement would still constitute an incomprehensible phenomenon. The fact that the arrow moves for a second presupposes that the same arrow flies for the first half-second, for the first half of this half-second, and so on, indefinitely. A. N. Whitehead remarked that Zeno must have been vaguely aware that if we consider the process in its entirety and ask ourselves *what it is* that has moved, it is impossible to give a certain answer. Whatever may have moved presupposes something that has moved in the preceding interval and, unable to complete the course, we are forced to conclude that *nothing* has actually moved.[5]

An issue implicit in the paradox of Achilles and the tortoise, and one debated from the earliest inquiries into analysis of the infinitesimal by Leibniz and Newton, relates to whether variables can reach their own limits. Newton in the *Principia* (Book I, Section I) defended the efficacy of the method of indivisibles, conceiving of quantities that were 'nascent' and 'evanescent', and argued that quantities and the relations between quantities that converge in a finite time in a continuous manner towards a relation of parity get closer to each other than any fixed distance, *in the end* becoming equal. Bishop Berkeley, in his celebrated polemical indictment of the mathematical use of the infinitesimal, asserted that mathematical science fails in the end to provide the clear and distinct ideas that many expect to eventually explain the mysteries of religion.[6]

Russell subsequently supplied a decisive contribution which indicated that Zeno's paradoxes regarding motion were far from mere sophisms. In 1903, in his *Principles of Mathematics* (par. 327) he would write:

> In this capricious world, nothing is more capricious than posthumous fame. One of the most notable victims of posterity's lack of judgement is the Eleatic Zeno. Having invented four arguments, all immeasurably subtle and profound, the grossness of subsequent philosophers pronounced him to be a mere ingenious juggler, and his arguments to be one and all sophisms. After two thousand years of continual refutation, these sophisms were reinstated, and made the foundation of a mathematical renaissance, by a German professor, who probably never dreamed of any connection between himself and Zeno. Weierstrass, by strictly banishing all infinitesimals, has at last shown that we live in an unchanging world, and that the arrow, at every moment of its flight, is truly at rest.[7]

Russell (*Principles*, par. 332) thought that the argument concerning the arrow articulated a fact that was simply elementary, and that the neglect of this fact had caused the philosophy of movement to be bogged down for centuries. His revisiting of Karl Weierstrass may be explained in the following way: along with Augustin-Louis Cauchy, Weierstrass was the first mathematician to clearly establish analysis without infinitesimals, asserting that a function $f(x)$ tends towards a limit L, for x that tends towards l, if, in correspondence with a given positive but nevertheless small value ε, it is possible to arrive at a positive number δ (dependent on ε) so that the distance of $f(x)$ from L is less than ε when the distance of x from l is less than δ. If $L = 0$, the function f approaches 0 as x tends towards l, but in the definition we deliberately avoid saying that the value of $f(x)$ becomes infinitesimal. Hence the idea of flow, of the dynamic tension of the variable towards its limit, vanishes, simply because the variables, within the confines designated by ε and by δ, do not move at all, assuming only the values that are proper to them. Immobility prevails over movement.

We can therefore define the *speed* of a body in an instant t only as the *limit* of the relationship between the distance covered and the time taken to cover it tending to the time variable at the instant t. This limit, a simple number, is the *derivative* of space as a function of travel time at the instant t. In this way, it was possible to avoid 'evanescent quantities' as conceived in the first developments of infinitesimal calculus. Thus Russell states in the *Principles* (par. 447):

> It is to be observed that, in consequence of the denial of the infinitesimal, and in consequence of the allied purely technical view of the derivative of a function, we must entirely reject the notion of a *state* of motion. Motion consists *merely* in the occupation of different places at different times, subject to continuity . . . There is no transition from place to place, no consecutive moment or consecutive position, no such thing as velocity except in the sense of a real number which is the limit of a certain set of quotients. The rejection of velocity and acceleration as physical facts (i.e. as properties belonging *at each instant* to a moving point, and not merely real numbers expressing limits of certain ratios) involves, as we shall see, some difficulties in the statement of the laws of motion; but the reform introduced by Weierstrass in the infinitesimal calculus has rendered this rejection imperative.

Thus rational and irrational numbers, conceived as limits of variables, inherited the actual and real nature of physical concepts such as velocity and acceleration. In those same numbers it is possible to recognize atomic entities that are points on a straight line. Movement could be interpreted, then, only through the coordinates of space-time, and thus by way of successive fixed and precise positions. 'Mechanics can only explain movement through immobility.'[8]

It was only with numbers – and this was the important conclusion – that the reality of the spatio-temporal continuum could be found. And the numbers that could perform this task could be either rational or irrational. In addition, the existence of real numbers (both rational and irrational) would appear, after Weierstrass, to be the effect of the free creation of a mathematician, albeit one induced by the objective properties of the numerical corpus. What better accord could there be between thought and nature, between freedom and the actual?

Whitehead explained that the objective world always expresses itself under the dual aspect of potential divisibility on the one hand and, on the other, of mutual relations and gradation that the process of division manifests, in every instant, as actual realities. The *perceived* world always appears in its potentially indefinite divisibility, while real, atomistic entities that define the *realitas objectiva* dwell in a system of mathematical relationships. As supporting evidence of his observations, Whitehead reiterated the thesis of William James, who when thinking about Zeno distinguished between the nature of a world immediately or actually perceived and the indefinite divisibility imagined by our reason:

> Either your experience is of no content, of no change, or it is of a perceptible amount of content or change. Your

> acquaintance with reality grows literally by buds or drops of perception. Intellectually and on reflection you can divide these into components, but as immediately given, they come totally or not at all.[9]

Whitehead remarked that 'continuity concerns that which is potential, whilst the actual is incurably atomistic',[10] but geometric continuity had already been conceived of, thanks to the theories of Cantor and Dedekind, as a domain of actual numbers. The arithmetical design of analysis had already *atomized* continuous extension. Actuality depends, in the end, in the theory of the numerical continuum, on defined numerical entities, constituent parts of a system of real divisions, of instantaneous events in relation to other events situated at some point on the continuum. Between numbers and points a correspondence is axiomatically established that is bi-univocal, and by way of numbers the points in space and instants in time acquire a new species of reality. As Russell writes (*Principles*, par. 326):

> In confining ourselves to the arithmetical continuum, we conflict in another way with common preconceptions. Of the arithmetical continuum, M. Poincaré justly remarks: 'The continuum thus conceived is nothing but a collection of individuals arranged in a certain order, infinite in number, it is true, but external to each other. This is not the ordinary conception, in which there is supposed to be, between the elements of the continuum, a sort of intimate bond which makes a whole of them, in which the point is not prior to the line, but the line to the point. Of the famous formula, the continuum is unity in multiplicity, the multiplicity alone subsists, the unity has disappeared.'[11]

For Poincaré the mathematical notion of the continuum is the product of the human mind, but it is also a kind of physical experiment that favours and renders its creation almost a necessity. The continuum's density appears as the consequence of a simple comparison between different measurements – as in the case, for instance, of the three weights A, B and C. It can happen that $A = B$ and $B = C$, due to the experimental impossibility of distinguishing A from B and B from C, but that we also have, experimentally, $A < C$. We would then be inclined to believe that *in reality*, contrary to the indications of the experiment itself, A is different from B, and B from C, and that other weights insert themselves between A and B as well as between B and C. This is a conclusion based on a principle of non-contradiction that does not depend on the precision of the instruments used to measure the weights.

In an analogous way, it seems necessary to impose the so-called completeness of the continuum, that is to say, the insertion into its framework of all the points which have co-ordinates that are rational numbers. There are in fact curves that intersect at points with coordinates that are *not* rational numbers. An example would be the point of intersection between the diagonal of a square and the inscribed circle. Assuming that only the points with rational co-ordinates exist, we will not be able to argue that this point of intersection exists in reality. If we draw a circle on the Cartesian plane, with its centre as the origin and its radius equal to the diagonal of a square of side 1, this intersects with the axis of the abscissa at a point the distance of which from the origin is measured by the irrational number $\sqrt{2}$.[12] Now it becomes impossible to suppress the tendency of our minds to consider the same point as equivalent to a real entity. Our mind cannot abide gaps in the continuum, and the most efficient way of avoiding them is to think of the coordinate *points* of intersection between lines as *numbers* that have the same *reality* as whole numbers and fractions. The number endows the point with reality, on the understanding that it is itself considered to be a real entity. As Russell explained, 'infinity and continuity appear together in pure arithmetic' (*Principles*, par. 435). It was this intellectual achievement that presented itself as a solution to the difficulty that Zeno raised regarding movement and the nature of the continuum. The modern solution of the Achilles paradox was based on assuming the reality or possibility of precisely that which Zeno considered paradoxical, that is to say, according to Russell's observation, the absence of a state of motion: a sacrifice that salvaged an indispensable fact: the actual existence of things. An actual entity, Whitehead observed, does not move: it is where it is, and it is that which it is.[13]

Russell had argued that the idea of a *state* of motion is not sound, because movement is made up of atomic positions occupied in determinate instants, both of these being equally accessible through real numbers, corresponding to points on a straight line. Aristotle had demonstrated (*Physics*, 234 a 24 ff.) that nothing can move in the instant (*nŷn*), and that because of this time is not made up of instants. Russell replied that, in effect, this is true, that nothing moves in the instant – and that this is compatible with a coherent theory of the arithmetical continuum supplied by the Euclidean metric, as had been elaborated by Weierstrass, by Dedekind and by Cantor. Only in this way could one guarantee the reality of that which changes and moves. The paradoxical becomes the real.

In Russell's commentary it is nevertheless possible to detect a kind of forcing, an effort to resolve a type of conflict between mathematical rigour and common sense. But the common sense, as also happened in the case of other mathematical theories, had to give way to a vision based on the evidence of formulas. Mathematics has always been an art of paradox, and its formulas have often provoked a reaction of incredulity in the very scientist responsible for having discovered or devised them. But mathematics is also an art of constructing, as far as possible, simulations and faithful models of our common conceptions, by means of definitions and theories capable of making us recognize what we expect. That quasi-imperceptible straining that we may discern in Russell's comments is followed by the explicit, unambiguous embarrassment of the commentary relating to Zeno's first paradox concerning motion by Hilbert and Bernays, and subsequently by Stephen Kleene:

> There is a much more radical solution to the paradox. This consists of recognizing that we are in no way obliged to believe that the mathematical representation of movement in terms of space and time is physically meaningful for intervals of space and time that are arbitrarily small; rather we have every reason to suppose that this mathematical model extrapolates facts from a certain domain of experience, that is to say movements within the orders of magnitude accessible until now to our perception, understood as a simple conceptual construction, analogous to the way in which the mechanics of the continuum effects an extrapolation in which one assumes that space is filled, in a continuous fashion, by matter . . . The situation is similar in all cases in which we believe that it's possible to exhibit an [actual] infinity directly as a fact of experience or of perception . . . A more careful examination shows then how an infinity is not

in any way given, but is interpolated or extrapolated by means of an intellectual procedure.[14]

But there was precisely no other path than that of extrapolating, of completing the facts derived from experience with a mathematical model of the continuum, which in turn could be reducible, as Hermann Weyl noted, to a mere symbolic construction. Aristotle (*Physics*, 263 a 25–30) had remarked that if the continuum is repeatedly divided into two halves, this cannot result in continuity with respect to line or movement. Movement, he emphasized, pertains most properly to the continuous, and in this there is undeniably an infinite number of halves – but only potentially, not in practice. Put simply, we could summarize it in the following manner: it is absurd to think that what moves does so by counting. But then it became clear that movement and the continuity of the straight line could not find an explanation through only the natural numbers, on the basis of which things would be enumerated one at a time. A new, more general theory of numbers would be necessary, as well as an extension of the idea of actuality or entelechy to encompass what at the end of the nineteenth century would come to be called, and not for nothing, the *real numbers*.

9. The Paradoxes of Plurality

Modern theories of the mathematical continuum have also sought to disentangle and clarify Zeno's paradoxes on the concept of plurality. According to the sources, the paradox on plurality had as a supposition the demonstration that a being without magnitude does not exist, and that in each thing that is endowed with magnitude and density its component parts are in some way distant from one another (29 B 1 DK). Now the parts that make up a plurality, consisting of an infinite number of elements, may have a positive magnitude or one that is null. In the first case, if the parts have an equal magnitude, however small, we obtain in any event an infinite magnitude. In the second case we have zero magnitude, because *whether part of a finite or an infinite total, the sum of individually null magnitudes is always null.* An infinite sum of points without extension must produce a combined extension of 0.

If by dividing a continuous line to infinity we reached elements that were indivisible, we would thereby encounter a paradox: if the indivisible elements have an equal length that is greater than 0, then the length of the line is infinite; if these elements have 0 length, their aggregate will also have a total length of 0. 'Hence, if there are many beings, it is necessary that they should be at the same time both large and small: large enough to have an infinite magnitude, and small enough to lack magnitude altogether' (29 B 1 DK).

In his polemic against the doctrine of Parmenides, Aristotle resorts to the thesis of Zeno according to which a point or a unit which is indivisible and without magnitude does not produce any effects whether it is added or subtracted, and consequently constitutes a non-entity, a pure nothing: 'If something is in itself indivisible, then according to Zeno's conception it is nothing. In effect, Zeno denies any existence to that which does not make a thing bigger or smaller by being added to or subtracted from it – assuming as self-evident the fact that what is real must have magnitude' (*Metaphysics*, 1001 b 7; 29 A 21 DK).

Hence, for Aristotle, Zeno argued that if something added does not produce increase it is *nothing*, and the same goes for something that by being subtracted does not result in a decrease. This is an important observation that may be connected to the Platonic idea of the nature of reality. For Plato, reality was founded on the *dýnamis*, the ability to produce an effect of any sort and, consequently, in particular, a growth or a diminution. At the heart of this conception was the analogy between the development of numerical progressions and the engine of growth inherent in *phýsis.* Similarly, in Zeno's reasoning, *the reality of a thing seems to depend on the capacity for producing growth or diminution.*

Aristotle seems to accept the vein of reasoning that refutes Parmenides, but declares himself in opposition to the thesis that a 0 degree of length is necessarily unreal: 'Something that is indivisible can certainly exist' (*Metaphysics*, 1001 b 14). In fact, Aristotle admits the existence of the geometric point, which he generally calls *stigmé*, a term that alludes to a kind of piercing, and hence to an actual insertion in space. Euclid, Archimedes and later authors tend instead to use the term *semeîon.*[1] Proclus would go on to find in these points a connotation of 'intelligible materiality', explaining that the point becomes materialized 'inasmuch as it appears in the recesses of the imagination' (*Commentary on Euclid's 'Elements'*,

96). Dianoetic intelligence assigns to the abstract unit a position in space, commutes it to a point, finding in the imagination (*phantasía*) an adequate figure, even if limited to a material and spurious existence. Augustine, too, would go on to register, in the *Soliloquies*, the difference between a real figure (an object of the intellect) and a figment created by the imagination, which the Greeks called *phantasía* or *phántasma.*

If Aristotle does attribute reality to a geometric point, he nevertheless denies that the point exists as something that is material, and when he refers to atomistic theories it is usually in order to refute them. The point may be defined as atomic because it is indivisible, but not because it is comparable to a corporeal atom, for according to Aristotle 'no body is an indivisible point' (*On the Heavens*, 296 a 17–18). Nevertheless, this does not imply for him that points, as for every other mathematical entity, have a separate reality from the perceptible one: 'If it happens that the entities which geometry deals with are sensory things, it does not study them as sensory objects, and mathematical sciences will not be sciences of the sensible; furthermore they will not be sciences of other objects separated from what can be perceived with the senses' (*Metaphysics*, 1078 a 1–5).

A point for Aristotle is distinguished from a whole (*monás*) only by its location in space. For Aristotle *the point is a unity that has position*, but the line is not made up of points, there is no continuity between two points and a point is not part of a line (*On Generation and Corruption*, 317 a 10; *Physics*, 215 b 19–20), just as in his view time is not made up of instants. The line, instead, is generated by the point by means of a movement (*On the Soul*, 409 a 4), and so the reality of the point resides once again in its capacity to produce magnitude by way of a kind of extension or growth.

What does it mean to say that a line is *made* of points, or *composed* of them? Zeno's argument is sustained by one single intuitive idea about composition, that of the summation or subtraction of elements, but the argument becomes more complicated if we take into account the different meaning more recent theories assign to terms such as sum, dimension, aggregate and quantity. Is the magnitude of the whole measured by its dimensions or by its cardinality (or power)? Dimension and cardinality do not depend on each other, because it is possible to conceive of linear combinations of points with 0 length but with the cardinality of the continuum. Cantor's *ternary* set has precisely these characteristics: it consists of all the numbers between 0 and 1 that have a representation in base 3 in which a number equal to 1 never appears (that is to say, the number is either 0 or 2). In geometrical terms, the whole is obtained by considering the interval [0, 1] and repeatedly cutting the first and the last third.

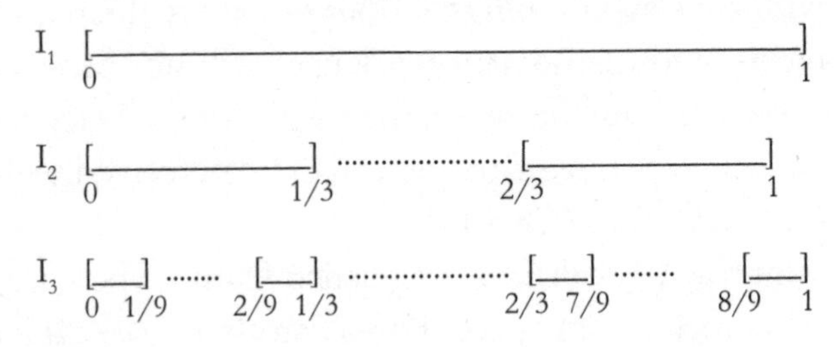

In this way we obtain an unlimited succession of subsets of the numbers included between 0 and 1, each of which consists of the union between disconnected closed intervals. Cantor's ternary set, which is the intersection of all of these wholes, *cannot be rendered in numbers*, because it is in bi-univocal correspondence with the set of real numbers which are greater than 0 and less than 1, and yet it is *discontinuous* and everywhere interrupted by gaps (the set of lacunae is everywhere dense, because between two of these, however close together, there is always another) and its length is null. Hence we obtain from this a combination of points in the interval [0, 1] that has the power of the continuum and a length equal to 0.

Nevertheless, it is possible that objects without dimension, such as the points of a line, can form a whole (the line) of superior dimension – that is to say, of dimension 1 – and the Euclidean metric defined coherently on the basis of the numerical continuum can assign a length greater than 0 to aggregates of individual points with no length. In this way, at an interval on the straight line between extremes A and B we assign a length that is precisely equal to $b - a$, where a and b are the real numbers corresponding to the points A and B. If a is different from b, this length is a number other than 0, even if the segment of the straight line traced by A and B is made up of points without extension. The metric theory of the numerical continuum, if taken to be coherent, would have among its consequences a confutation of the reasoning that underpins Zeno's paradox on the question of plurality.

Even if we hypothetically propose that the universe is made up of contiguous bodies, Aristotle noted (*On Generation and Corruption*, 325 a), if the universe is completely divisible, there can be neither unity nor multiplicity, everything is void – and, moreover, we cannot know how to contend that pieces of extension exist that are indivisible; they would be pure fictions. This risk of facing the fictive void or the inconsistency of conventional assumptions also helps to explain the attraction of numbers in antiquity as the foundation of the real. Since the ancient atomistic theories, albeit not accepted by Aristotle, the escape route from Zeno's paradoxes has focused on a possible element of reality that consists in numbers and their relations, and in their capacity to measure geometric shapes, the relations of equivalence and similitude. The problematic tension between arithmetic and geometry, between numbers and points on a straight line, reveals all the difficulties that are inherent in this programme. The search for actualities to stand against the indeterminacy of a regression or a progression *ad infinitum* must have been, in particular, a motivating force behind the atomistic theories developed by Leucippus and Democritus. Democritus had propounded the qualities of objects as being the product of custom and convention: 'Colour is according to convention, sweetness is down to convention, bitterness is according to convention – while the only real things [*etehêi*] are atoms and the void' (68 B 125 DK). Despite this, Democritus also contended that 'we know nothing of the real: in effect, the truth is to be found in the depths' (68 B 117 DK).

Later on, Lucretius used the same argument about dichotomy as Zeno in order to demonstrate that the existence of atoms is the only alternative to the complete dissolution of the real (*On the Nature of the Universe*, I, 615–27):

> Besides, unless there is some smallest thing,
> The tiniest body will consist of infinite parts,
> Since these can be halved, and their halves halved again,
> Forever, with no end to the division.
> So then what difference will there be between
> The sum of all things and the least of things?
> There will be none at all. For though the sum of things
> Will be completely infinite, the smallest bodies
> Will equally consist of infinite parts.
> But since true reasoning protests against this,
> And tells us that the mind cannot believe it,
> You must admit defeat, and recognize

That things exist which have no parts at all,
Themselves being smallest. And since these exist
You must admit that the atoms they compose
Are themselves also solid and everlasting.

Contrary to a still-widespread belief, mathematics is *not* born out of pure abstraction, as an abstract model of physical reality. The principle of reality had to be entrusted, initially, to a sort of numerical atomism, in opposition to the irreality evoked by the unlimited divisibility of the continuum. Numbers, relations and algorithms originate in contradistinction to the *ápeiron*, to the infinitely big and the infinitely small. This is evident in the case of the Pythagoreans: their numbers were conceived of as ordered sequences of points separated in space, in a way that seems to be responding precisely in order to found a universe of existing entities existing in a way that is actual and effective; not an abstraction, as at first it might appear, but on the contrary an effective realization of the substance of things.

According to Aristotle (*Metaphysics*, 1080 b 16 and 1036 b 8; 58 B 9 and 58 B 25 DK), the Pythagoreans maintained that all sensitive substances were made of numbers, that the entire universe was made up of them, and that geometrical shapes such as the circle and the triangle should be regarded in the same way as the flesh and bones of humanity, and as the bronze or stone of a statue. Plato himself, Aristotle reminds us (*Metaphysics*, 987 b 22; 58 B 13 DK), thought that numbers were the cause of the substance of things, and in fact equivalent to the things themselves. Leucippus and Democritus would have followed this line of reasoning and, indeed, Aristotle helpfully points out that the atomists thought, just like the Pythagoreans, that 'everything that exists in the universe is made up of numbers, or develops from numbers' (*On the Heavens*, 303 a 8–9). The immanent element that was constitutive (*stoicheîon*) of numbers was unity, just as the point was constitutive of the straight line. The one and the point are small, simple, indivisible, everywhere manifest, a circumstance that induced ancient philosophers to treat them as principles (*Metaphysics*, 1014 b).

This programme of the Pythagoreans and the atomists was perfected by the mathematics of the nineteenth century, when Weierstrass, Cantor, Dedekind and others developed a coherent theory of the numerical continuum, defined by the field that includes rational and irrational numbers. We can therefore understand just how important it was to establish that irrational numbers *exist* and that they have the same ontological status as natural whole numbers.

10. The Limited and the Limitless: Incommensurability and Algorithms

Why do whole numbers and algorithms, unlike irrational numbers and the concept of the continuum, have their own unquestionable reality? If we follow the Platonic criterion for defining what is real and what is not, we would have to revisit the passage in *Philebus* (16 c) in which it is asserted, thanks to knowledge inherited from the ancients, who were nearer to the gods, that 'the realities that we claim always exist, given that they are constituted by the one and the many, have interrelated within themselves the limited and the limitless'. This offered a vision not far removed from that of Parmenides, who asserted, according to Proclus, that 'being is one only as far as the concept [*eîdos*] is concerned, and multiple instead as far as the evidence of our sensory experience goes [*enárgeia*]' (29 A 15 DK).

For Plato the natural whole number, the *arithmós*, is precisely that which comes between (*metaxý*) the infinite and the one (*Philebus*, 16 d–e), between the limit and the limitless, that which mediates between the former and the latter. With algorithms, one was in a position to calculate, that is to say, to *produce* numbers and relations between numbers, and this production was realized between two opposite poles, which were in effect the one, conceived as the seminal *lógos*, and the limitless, the indefinite growth of numbers and of geometric shapes in different sizes. Number and relations were generated by the unit in sequences that could be extended infinitely, by way of successive operations of division,[1] thereby forming a dense intermediate texture that constituted *reality*, or an adequate representation of it. To approximate the entities that we now call irrational numbers there already existed in antiquity what Cantor would go on to term *fundamental sequences*, sequences of fractions converging at a point on the real straight line. But for the Greeks reality did not consist of that which Dedekind would come to call a 'section' or 'cut' of the rational corpus, or in the number with infinite decimal digits that we denote, for example, with the symbol $\sqrt{2}$. That was an inexpressible lacuna, not an entity to which an actual physical existence could be attributed. Hence they lacked a precise idea of numerical approximation: there was no entity to approximate, and consequently it wasn't even conceivable that there was a distance between the fraction and the limit of the sequence. The reality of the relations between geometrical magnitudes that were incommensurable, such as the diagonal and the side of a square, resided entirely in the relations between whole numbers, in the *lógoi* effectively calculated, and in the law that allowed them to be calculated. However, this circumstance in some cases did not impede the possibility of associating to a geometric line a whole number capable of measuring it approximately, even in the case when the line, for instance, could be the diagonal of a square with side 1. For Plato (*Republic*, 546 c) the *rational diagonal* of 5, that is to say $7 = \sqrt{49}$, could be compared to $\sqrt{50}$, that is to say, the exact length of the diagonal of the square with a side 5. The number $7 = \sqrt{49}$ is a whole number, while $\sqrt{50}$, which is an irrational number, constituted a lacuna for the Greeks, an entity that cannot be represented by means of integers.

Plato's argument is based on Pythagoras' theorem applied to a right-angled triangle with both sides of length 5. Because the sum of the squares on these sides is equal to the square on the hypotenuse, from the equivalence $5^2 + 5^2 = 50$ it follows that the length of the hypotenuse is $\sqrt{50}$. The square root of 49 is an approximation of it, which has the advantage of coinciding with a natural whole number, 7, instead of an irrational number such as $\sqrt{50}$. The constructions using ropes and wooden pegs on the Vedic fire altars as described in the *Śulvasūtra* make use of a similar approximation, that is to say: $5^2 + 5^2 \approx 49$.[2] In any event, the measurement of the length of all bodies is never exact, and normally consists, as the text has it, of something only 'more or less' correct.[3]

The concept of approximation, already foreshadowed in the numerical procedures of antiquity, would eventually become one of the cornerstones of mathematical thought. Since time immemorial (with clear allusions to the fact, for example, in the Vedic texts) it was known that quantities cannot be measured exactly, but only through approximations, by excess and defect (exceeding and falling short of the mark).

The question of how to approximate irrational numbers did not merely relate to applied aspects of calculation. The techniques for measuring degrees of approximation would become the very same techniques of mathematical thought. In fact, the Euclidean definition of relation is indebted to prior techniques of numerical approximation by excess and defect; the definition of limit we owe to Weierstrass is in turn indebted for its logic to the existence of numerical algorithms (in which the same symbols ε and δ appear that in previous times had denoted the *error* of approximation); the analytical formulas used for approximating a function serve to define general concepts, such as the stability of a system of calculation; a theorem demonstrated in 1844 by Joseph Liouville (see p. 158 below) established the degree of approximation that was attainable by approximating the roots of an algebraic equation, and even allows the quantification of a crucial phenomenon in automatic calculation, that is to say, the speed of *growth* of the numbers calculated.

We can venture a few conjectures regarding the tools that made it possible, before Euclid, to determine the existence of incommensurable geometric magnitudes. We should point out immediately that different *demonstrations* of the incommensurability of the diagonal and the side of a square were possible, and various *procedures* for calculating rational approximations of $\sqrt{2}$. But we invariably find in these the theme of the growth and decrease of quantities. In one of the possible demonstrations the same dichotomous criterion used by Zeno appears, even if Aristotle seems to cast doubt on the possibility of such an alignment between incommensurability and the generative techniques behind the paradoxes (*Prior Analytics*, 65 b 16–21). The dichotomous method in any case leads to the construction of a limitless succession of squares, and in this way the age-old paradigm was repeated: building from an initial nucleus, formed by a geometric figure, to an infinite number of shapes similar to the first. Let's assume, by reasoning *ad absurdum*, that the diagonal and side are commensurable, so that we may construct a sequence of squares of decreasing scale, by means of progressive halvings, associating each time with the diagonal and the side, as a consequence of the initial hypothesis of commensurability, a positive whole number. The linear segments become progressively smaller, as do the corresponding numbers. But the sequence of diagonals and sides is limitless,

Figure 5

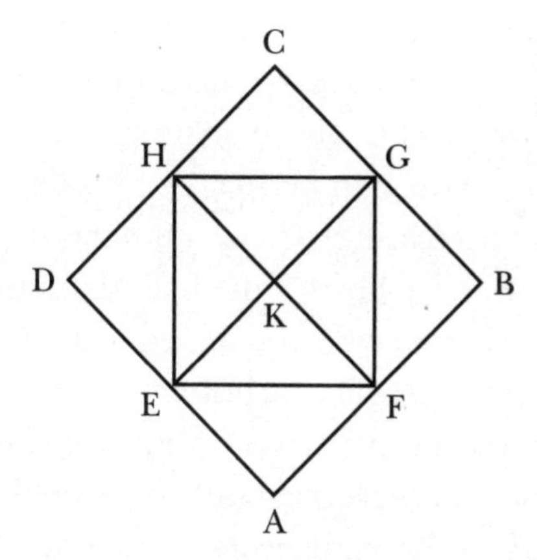

whereas the decreasing size of the numbers in the sequence cannot be: hence the absurdity.[4]

If we suppose, absurdly (Fig. 5), that the diagonal EG of the square EFGH is commensurable with the side EF (in the figure, E, F, G and H are intermediate points of the sides of the square ABCD), it should be possible to express the relation EG:EF as a relation between two whole numbers d and s (Euclid). Now, the square ABCD and the square EFGH represent, respectively, d^2 and s^2 – one of which, as shown in the diagram, is the double of the other. Therefore, d^2 is an even number and, given that the square of an odd number is odd, AB is also even. This implies that its half AF represents a whole number. Now EF represents a whole number, and because of this the relation between EF and AF is a relation between whole numbers. Consequently, we have constructed a square EKFA, the side of which is half of the side of the square ABCD, so that its diameter and its side represent whole numbers. The construction may be repeated indefinitely, generating squares that are progressively smaller, with the side each time being half the size of the preceding one but always equal to a whole number. The numbers become increasingly small, just as the squares do, but while the squares, which comprise continuous magnitudes, may be reduced indefinitely, the whole numbers cannot become inferior to 1. The indefinite decreasing of the numbers is incompatible with the indefinite halving of the lines, because the process should necessarily conclude in a unified segment or one measured by an odd number. The dichotomous method thus offers a perfect example of that kind of incompatibility between numbers and geometrical diagrams that would go on to lead to the discovery of incommensurable quantities.

The dichotomous method was not the only possible expedient device for studying incommensurability. Another demonstration with some affinity to the preceding one, which could even be placed at the origin of the study of incommensurability, was strictly tied to the algorithm that makes it possible to approximate, up to the required precision, the relation between the diagonal and the side of a square. The two procedures, demonstration and algorithm, are like two sides of the same coin – each is the reverse of the other – and can still be illustrated with the construction of an indefinite number of squares. The basis of the demonstration procedure is *antanaíresis* or *anthyphaíresis*, the process of successive subtractions known as the 'Euclidean algorithm', the technique for finding the largest common denominator between two magnitudes with the calculation of quotients and remainders. Given a square ABCD (Fig. 6), one marks on its diagonal BD = d a segment BE equal to its side l. This operation is nothing other than the division of d by l, the result of which is 1 with a remainder equal to ED. One takes therefore the remainder ED as the side l' of a subsequent smaller square, smaller than

Figure 6

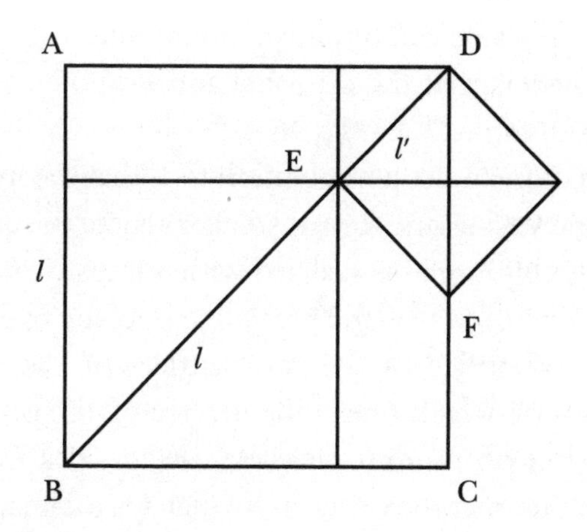

half of d, and proceeds to construct from the square of side l' and diagonal $d' = \mathrm{DF} = l$ a third square with the same procedure. The construction can be repeated indefinitely, and in every square Q the diagonal d is equal to the sum of the side plus the side of the subsequent smaller square Q′, that is $d = l + l'$.[5]

The reasoning develops by absurdity. If the diagonal BD and the side BC of the larger square are commensurable, then both are exact multiples of the same unit of measurement U. So it also follows that the difference BD − BC = EF and the difference DC − FC = DC − EF = DF are both multiples of the same unit of measurement U (because they are differences between multiples of U). That is to say that the side and diagonal of the smaller square are commensurable and exact multiples of U. Constructing ever smaller squares, every diagonal and every side becomes a multiple of U. But these multiples are always smaller: if it were BC = 40U, for example, then it would be a maximum of EF = 39U. Every side would be equal to nU, with n always being a smaller positive whole number. But n cannot become smaller than 1, whereas the squares can diminish indefinitely. Consequently, the conclusion is that the diagonal and side of a square are *not* commensurable (*Elements*, X, 2).[6]

Now what one must pay attention to is that the inverse trajectory, in increasing *scale* from a smaller square to squares that are progressively larger, is realized with a transformation that is precisely the reverse of the preceding one: the two transformations are identified by the two matrices of two rows and two columns of which one is the inverse of the other.[7] With the succession of progressively increasing sides l and diagonals d we construct the relations $d{:}l$ that approximate $\sqrt{2}$. We thereby obtain, that is, an *algorithm*, a process of calculation of rational numbers (the sequence of fractions d/l) that, as l increases, that is to say as $1/l$ decreases, gets progressively closer to $\sqrt{2}$ (without ever reaching it). It is the same algorithm of lateral and diagonal numbers that Theon of Smyrna (first to second century AD) refers to in his *Expositio rerum mathematicarum ad legendum Platonem utilum*. A clear demonstration of the co-presence of the 'big' and the 'small' as the Platonic designation of the *ápeiron*: *the numbers d and l become increasingly bigger while the unit of measurement* $1/l$ *becomes increasingly smaller*. Also, the *distance* of the relation $d{:}l$ from $\sqrt{2}$ becomes increasingly smaller with the growth of l, but it can never be 0, and this could very well have represented a revealing argument for understanding that the diagonal and the side of a square are incommensurable.[8]

We can assert that 2 is precisely the *limit* of the succession of relations $d^2{:}l^2$. In fact, we can calculate l in such a way that the distance between 2 and the relation $d^2{:}l^2$ is less than an arbitrarily small quantity ε, because the distance is equal to $d^2{:}l^2 - 2 = \pm 1/l^2$. This means that the succession of relations becomes increasingly dense in proximity to 2: when a fraction ε which is arbitrarily small is fixed (for example $\varepsilon = 1/n$, with n equal to a very big number), there exists an *infinite* number of relations $d^2:l^2$ that are distant from 2 by less than ε, while only a finite number of relations are distant from 2 by *more than* ε. In terms of fractions, we would say that the sequence of fractions d^2/l^2 *converges* at 2, or that d/l *converges* at $\sqrt{2}$.

This way of conceiving the limit, alongside the modern formulation of Cauchy and Weierstrass, is in a certain sense counter-intuitive, but it answers to a more precise mathematical criterion. In the case of the succession of relations or fractions d/l we could now repeat what has already been stated in relation to the limit of a function. Intuitively, we would be inclined to regard the unfolding of the sequence of fractions in a dynamic sense, as if it were a question of the movement of a fraction that tends to get closer to a point of destination or target. But the dynamic unfolding, as it was conceived by the majority in the analysis of the infinitesimal in the seventeenth century, does not find a clear formulation in mathematics and leaves equivocal gaps regarding the effective attainment of the limit. The most rigorous definition overturns the perspective here: we *first* select an area that is arbitrarily small for the limit L (of radius ε) and determine on the basis of this from which point onwards the generic end of the succession is definitively in that area. The same conception of the limit according to Cauchy and Weierstrass, shaped by an idea opposed to that of a dynamic tension, is aligned with a static and atomistic conception of numbers and correlates to an idea of the infinite understood as actual rather than potential.

It is perhaps worth pointing out that this overturning of perspective afforded by the modern definition of the limit of a sequence closely follows the logic of approximate calculus. In the algebraic *computatio* that preceded Cauchy and Weierstrass the symbol ε typically stood for the residual error in the calculation of an irrational number by means of a sequence of fractions stopped at a certain endpoint (at a distance ε from the solution). The error ε could be pre-assigned in such a way as to fix a criterion determining the endpoint. Similarly, in the sequence of fractions d^2/l^2 we can decide to arrest the procedure when the residual error ($1/l^2$) is less than a pre-set fraction ε.[9]

The sequence of relations $d^2:l^2$ defines a sequence of intervals $I_1, I_2, \ldots I_k, \ldots$, each one of which includes the number 2 and has as its extremities one approximation by defect (rounding down) and one by excess (rounding up). The intervals are included one within the other and become progressively smaller around 2, but their length never becomes null. A similar conclusion obtains for the relations $d:l$. But in this case the intervals include an entity of a kind different from 2, not a whole number but a number the square of which is equal to 2, designated by the symbol $\sqrt{2}$. Now it becomes impossible to avoid the perception of the actual presence of this entity, even though irrational, but one is obliged to consider as really existing, in opposition to the indefiniteness of the *ápeiron*, only the relations $d:l$. One may then define the number $\sqrt{2}$ as the same sequence of infinite intervals that include it, because the extremities of these intervals are the rational numbers that are actually capable of being calculated and that are precisely equal to the relations $d:l$. In such a case we may be inclined to substitute an individual entity for the graduality of approximation, a symbol that corresponds to a new number and denotes the sequence of intervals.

If we now think of any irrational number as an indefinite

sequence of intervals that includes it, and we denote it with a symbol, we are led to conceive of a new domain of symbol-numbers as a multiplicity of infinite separate individuals. This is a decisive step, facilitated by the possibility of formally defining an arithmetic, among these symbol-numbers, similar to that of rational numbers.

The following is also a decisive factor: the indefinite sequence of intervals included one within the other gives way to a *separation* of the set of all the rational numbers into two classes. The first class contains all the rational numbers that are on the left-hand side of the intervals I_k, for a k that is sufficiently large, that is to say, which are less than the extreme left of all the intervals I_k, with the exception of a finite number of them. The second class contains rational numbers that are superior to the right-hand extremity of all the intervals I_k, with the exception of a finite number of these. To these two classes we add a third class formed by the rational numbers contained within all the intervals. Now the third class, if it is not empty, contains a single rational number r. If the third class is empty, that is to say, does not contain any rational number, the sequence of intervals *represents* an irrational number.

This division of the set of rational numbers is precisely what Dedekind, at the end of the nineteenth century, would define as a *section* of the rational corpus. The *real* numbers, the rational and the irrational, would then be conceived as sections. But the concept of section is already implicit in Greek mathematics, precisely in the mechanism of *antanaíresis* and in the construction of the ratios d/l. The reasoning is similar to that proposed by Poincaré in *Science and Hypothesis* (1902): we almost feel obliged to conceive of a number, albeit irrational, exactly in the same way that we feel obliged to recognize as existent and real the point of intersection P between two curves on the

Cartesian plane, getting closer from right to left through the rational abscissas increasingly close to the abscissa of *P*.

It may appear surprising to find, in the *same* recursive formula with which we calculate the lateral and diagonal numbers, the simultaneous presence of a demonstration of non-existence (a common measure *m* does not exist) and the effective construction of rational approximations of that which, in the rational field, cannot exist. These are two aspects of mathematics distinguished by Proclus: one theoretical in kind, the other of a problematical type. With the first, one attempts to establish a theoretical result, such as the existence or non-existence of a rational solution; with the second, one effectively seeks to construct the solution or a fraction that approximates it, if it does not exist. These are two forms of complementary reasoning that manifest themselves in the very same formulas.

The construction of increasingly large or increasingly small squares is a process without limit. Nevertheless, even in the absence of the last step, the form and regularity of the procedure allow us to reach a conclusive thesis regarding the entire succession of squares and an *infinity* of possible choices. Implicit within this there is what in logic is called the universal quantifier: *any* segment, however small, cannot be a common measure of the diagonal and the side of a square. We have a *demonstration of non-existence*, but at the same time we also touch a reality, by way of a geometrical proof. *Reality* and *existence* do not coincide; they are two distinct concepts that operate on different planes.[10] Reality is something that is asserted by necessity, not on the basis of any kind of convention. Such conventionality would falsify its nature, weakening that aspect of inevitability and injunction that is imposed by the very formulas and the algorithms themselves. We do not determine anything; it is the numbers that arrange themselves according to their own law, which is ultimately independent of us.

The succession of sides l and of the diagonals d of squares, taken in ascending scale, is infinite – but the succession of the relations $d{:}l$ has a *limit*, represented by what we call 'the square root of 2', because the relation $d^2{:}l^2$ converges at 2.[11] The concepts of relation and limit, joined to the calculation of approximations whose distance from the limit progressively decreases, allow us to encompass the inconclusiveness of the *ápeiron*. Even if it is open and potential, the *ápeiron* is, so to speak, perfected by the limit, and obeys a regularity of development from which *final* conclusions are reached. In the mathematical concept of relation we find the antidote to the indefiniteness of the *ápeiron*, and in the progressive generation of numbers d and l it is possible to 'grasp the limitless at a stroke, because the limitless is limited by a relation that is grasped in a single action'.[12] This is precisely what Plato says in *Philebus* (16 c): an entity that is knowable and real is a combination of the limited and the limitless.

In Iamblichus, in Theon of Smyrna and in Proclus, that is to say in the original sources that go deepest into the detail of algorithms, there is no discernible explicit reference to the fact that the relations $d^2{:}l^2$ approximate 2. It has been pointed out that Plato alludes to the 'rational diagonal of 5' (literally, the effable and expressible diagonal), that is to say, the number 7, which we know to be an approximation of the diagonal of the square with a side 5, which is on the contrary irrational or, better still, ineffable and inexpressible. But we do not encounter the concept of approximation if not in the contrast between rational and irrational. Only in a passage of the *Statesman* (287 c) does the Stranger from Elea refer to *diaíresis*, to the strategy of searching for successive divisions, saying that 'one must always divide, as far as possible, according to the number closest to two' – almost in the same way as a 'sacrificial victim' is dissected. This *proximity* (in Greek, *engýtes*) may allude to the narrowing of distances, to the dense accretion of relations that tend towards irrational numbers. The irrational numbers such as $\sqrt{2}$ will be defined formally, in the nineteenth century, as sequences of rational numbers p/q (where p and q are whole numbers) that become denser the closer they get to their limit.

Interpreted as the 'rational diagonal of 5', the number 7 alludes to the possibility of expressing the irrational through the rational by means of numbers. In the relations between numbers (positive whole numbers) we find the real, concrete and actualized expression of that which we will only be able to express in negative terms, since the relation between magnitudes for a common unit of measurement *does not exist.* This is a circumstance that is reinforced with the comment by Theon of Smyrna whereby the unity (*monás*), which is the initial value of both side l and diagonal d, on the basis of which the relations $d{:}l$ are constructed, 'as the principle of all things, must have the potentiality [*dynámei*] of both the side and the diagonal'.[13] In the *dýnamis*, as Plato insisted (*Epinomis*, 990 c), we find the *reality* of numerical entities with which it was possible to express the inexpressible.

The salient fact, already evident in Greek mathematics and Vedic calculations, will be that relations of equivalence and the operation of enlargement or reduction of *geometrical* figures are generally in close relation to analogous recursive operations on *numerical* magnitudes. These operations make the whole numbers p and q that define the fractions p/q grow, and the fractions approximate the relations between incommensurable magnitudes; and the incommensurability is visible, in turn, in the behaviour at the limit of sequences of similar figures that grow or decrease.

Now, it is well known that the approximation improves with the growth of the numerator p and the denominator q of a fraction, a characteristic that is already traceable in the lateral and diagonal numbers that increase with a recursive law. But a calculator, whether human or digital, always operates with a number t prefixed to the numerals and rounds the numbers with more than t figures, causing irreparable losses of information as it nears the limit. For this and other reasons, the digital calculator prefers to represent the fractions p/q as 'floating points'.[14] But this expedient is not enough, because as we shall see the growth of numbers can compromise the stability[15] of the procedures, which is a fundamental requirement for their effectiveness.

Incommensurability, as we have seen, has to do with the structure of the continuum. The central question is always whether the continuum is made up of indivisible parts, or of parts that in turn may be divided indefinitely. In a passage of the treatise *De lineis insecabilibus* ('On Indivisible Lines'), included in the *Corpus Aristotelicum* (970 a 14 ff.), the inconsistency of the concept of the indivisible line is demonstrated. One envisages a succession of squares regulated by *antanaíresis*, the Euclidean algorithm for the calculation of the highest common denominator of diagonal and side. If one were to suppose, absurdly, that there was an indivisible line, this would be the side of a square the diagonal of which, divided by the side, would as a result give a line with a length inferior to that of the side – otherwise, the square constructed on the diagonal would be at least four times rather than just twice as large as the former. One would then find a line smaller than the side, assumed to be indivisible, hence the absurdity. This is one reason why Aristotle could assert that atomistic theories are unsupported by anything mathematicians are able to demonstrate, including, one imagines, the incommensurability of the diagonal and the side of a square.

The possibility of representing the numbers d and l by means of unlimited sequences of similar figures is signalled throughout the extant literature, albeit with differing emphases and geometrical implications. The formulas for generating the numbers d and l seem connected to a variant of the algebraic formula which corresponds to a theorem of Euclid's (*Elements*, II, 10), from which it is possible to extract a law of construction relating to the succession of squares. This demonstrates that in the *Elements* we find not only, according to a dubious and widely discussed thesis, a *geometrical algebra*, that is to say, an algebra couched in geometrical language, but also a *computatio geometrica*, a numerical calculus the structure of which is correlated to the properties of geometrical figures. One needs to distinguish between these two aspects because the algebraic formulas are *not* algorithms. For example, the formula that denotes the product of three numbers, $a \times b \times c$, corresponds to *two* possible procedures, suggested by the two respective expressions $(a \times b) \times c$ and $a \times (b \times c)$, which can give two completely different results.[16]

There is in the end an essential nexus that links to Vedic geometry the demonstration of incommensurability realized by means of construction indicated by Figure 5. In Figure 5 there are four equal squares present, namely CHKG, BFKG, KFAE and HDEK. Now the bigger square composed of these four squares coincides with the body of one of the principal Vedic altars, as indicated in Figure 1. Tracing the diagonals in each one of the squares of the central body of the altar, and repeating the operation in order to realize a progressive *emboîtement*, or nesting, of geometrical figures, one ultimately demonstrates the incommensurability of the side and the diagonal of a square. The same construction also leads to the intuitive, immediate demonstration of Pythagoras' theorem applied to equilateral, right-angled triangles. It is presumed that this demonstration was known to the authors of the Vedic treatises on the construction of fire altars. Albert Bürk, in his commentary on Āpastamba's *Shulba Sūtra* (*Śulvasūtra*), does not fail to point out that the central body of Agni is none other than *ātman*, the all-knowing generator of the world, the central nucleus of every living being and universal consciousness, which remains, beyond growth and the multiplicity of forms, always equal to and the same as itself.[17]

11. The Reality of Numbers: Cantor's Fundamental Sequences

Mathematical theories of the arithmetical continuum developed at the end of the twentieth century seem to be capable of confuting Zeno's paradox on the nature of plurality, which relies on the presupposition that a finite or infinite *sum of non-extended magnitudes*, such as points on a straight line, *is always a null magnitude*. If this presupposition were verified, then reality would vanish into the void and into non-being: an unacceptable thesis that, in Eleatic logic, imposed the conclusion that multiplicity does not exist and that being is *one*. Parmenides' thesis that Being is one, unmoving, homogeneous, isotropic and continuous, and that every multiplicity is illusory, was still less paradoxical than the consequences that follow from the existence of any plurality.

The theory of the numerical continuum is based above all on the bi-univocal correspondence between points on a straight line and real numbers (rational numbers, or fractions, plus irrational numbers) that allows us to consider the *real straight line* as both a geometrical and a numerical entity. This entity is definable in different, equivalent ways. A real *number*, which is also a *point* on a straight line, is an infinite decimal number that is either periodic or aperiodic. In an equivalent way the real numbers consist of the so-called *fundamental sequences* of rational numbers (fractions). Their ontological status is in large part revealed in Cantor's own words:

There is a first generalization of the notion of numerical magnitude in the case where in virtue of a law an infinite series of rational numbers is obtained $a_1, a_2, \ldots, a_n, \ldots$, such that the difference $a_{n+m} - a_n$ becomes infinitely small [in absolute value] as n increases, whatever the value of the positive integer m, or in other terms, such that, once assigned an arbitrary positive rational number ε, an integer n_1 exists for which $a_{n+m} - a_n$ [in absolute value] is less than ε when $n \geq n_1$, and m is an arbitrarily chosen positive integer. I express this property of the series as follows: 'the series has a determined limit *b*'. Hence these words serve only to enunciate this property of the series, without for the moment alluding to another; and as we connect the series to a particular sign *b*, we must in the same way give different signs b, b', b'' to different series of the same species.[1]

For Cantor, the real numbers coincide with the symbols that denote sequences of fractions that become infinitely denser in proximity to the limit, and are such that in the case of indices (a simpler way of expressing large numbers) greater than a value n_1, *of which for the moment the actual calculability is not required*, the difference in absolute value between two fractions is less than any positive rational number ε assigned. Discernible in this explanation is the essentially formalist spirit with which Cantor introduces the real numbers that consist precisely of these sequences. The *sign b* denotes the sequence and consequently also a *number* (rational or irrational), though it is only a symbol introduced to represent the series in a concise manner.

Hence the real numbers *are*, by definition, fundamental sequences or successions, and their arithmetic is a formal extension of the ordinary operations of the same sequences.

The sequence of relations between diagonal and lateral numbers is fundamental in Cantor's terms and identifies the real number designated with the symbol $\sqrt{2}$. The algorithm of approximation of $\sqrt{2}$ devised by the Pythagoreans was therefore a kind of *effective realization* of the non-extended point corresponding to the $\sqrt{2}$.

Now Cantor demonstrated that any collection of intervals of the real straight line that are not null and do not overlap is a set that is at most numerable, that is to say, is in one-to-one correspondence with the set of positive whole numbers. Instead the agglomeration of real numbers conceived as sequences of intervals of a length tending progressively towards 0 forms a set that is *not* numerable. Cantor showed this with the famous diagonal argument.

Cantor's theory is not based on the indefinite divisibility of a straight line that tends, *in the end*, to points without extension, but on the actual pre-existence of a non-numerable infinity of points/numbers to be thought of as successions of intervals, the length of which progressively diminishes and tends towards 0. If we formally associate a symbol with each one of these successions (as Cantor proposed, referring to fundamental sequences) we can think of the continuation of these individual symbols as an aggregate of atomistic points in the context of which we can define intervals of positive *length*. Ultimately, we are thus dealing with a symbolic construction, of a mathematical *model* of the continuum that, while allowing us to build an analysis, does not wholly correspond to our intuition. But now the set of real numbers between the two extremes a and b, where $a < b$, or the interval $[a, b]$, has a length defined precisely by the real number $b - a$, which is not equal to 0. Hence the theory of Cantor and of Dedekind, if it is coherent, virtually extricates us from Zeno's paradox because it allows us to define intervals of a length which is *finite and not null* and formed by an innumerable collection of non-extended points.

In this way we can succeed in giving to the continuum a connotation of actuality and efficacy, if we are to credit Whitehead's thesis that actuality has a strictly atomistic character. Mathematics, moreover, is the essential tool for understanding whether this idea of atomistic realism can result in an acceptable and coherent theory. As John Locke noted (*An Essay Concerning Human Understanding*, II, 17, par. 9), we find in numbers the most adequate concept for understanding the nature of the infinite, and for getting to grips with that confused accumulation of elements into which the infinite drags us, and within which the mind, without other sufficient means, ends up by losing itself. Physics also studies atoms. But physics and mathematics pursue different aims in this field: if the physicist studies the reciprocal relation between thought and reality, the mathematician instead seeks to clarify the nexus between thought and formulas, and as a result is oriented towards the elaboration of appropriate and sufficiently plausible models of our perceptions, a necessary prerequisite for them to be applied to the real world. Despite these differing orientations of inquiry, however, both physics and mathematics seek to establish the reality of the world.

We are left, however, with various problems relating to this matter. A real number is an element of an abstract domain the axioms of which are satisfied by a variety of mathematical entities that may not be numbers. We can therefore seek to define the *essence* of mathematical entities only by means of an isomorphism, with a lack of categoricalness almost impossible to eliminate.[2] Moreover, we are compelled to think of a point on the number line as an indivisible entity. Can we then adapt this concept of number as a fundamental sequence, as a section or rather as a sequence of intervals? Both Dedekind and Cantor felt obliged to *create* in some way a new entity that would correspond to a sequence or to a section, and that could be designated with a symbol to which we could assign as much reality as to any whole number.[3]

For transfinite numbers (larger than all finite numbers but not necessarily absolutely infinite) Cantor fixed an ontological status akin to that of real numbers, on the basis of a formal definition of both. It is no coincidence, then, that he should insist on different occasions on the inherent *reality* of numbers of whatever kind, because these numbers, however abstract, were the basis of an actuality of an atomistic kind – the only one that seemed to lend itself to being juxtaposed, with respect to the axioms of a complete and ordered numerical field, to the merely potential infinity of an iterated division of the continuum. For this reason also, and not only because it was necessary to define transfinite numbers in a coherent way, Cantor was inclined to consider the actual infinite to be realizable in mathematics.

Cantor accompanied the announcement of his discoveries with the declaration that '*the new numbers obtained in this way always have the same concrete precision and the same objective reality as the preceding ones*'.[4] Whole numbers for Cantor are already *actual* entities, in the sense that, on the basis of their definition, they have a position which is perfectly determined and distinct in our thought, to the extent of determining and modifying its development. The *actuality* of numbers means that they 'must be considered as an expression or image [*Abbild*] of processes and relations in the *external* world, distinct from the intellect'.[5]

This dates back to a monograph published separately in 1883, *Fundamentals of a General Theory of Multiplicities*, the first public advocacy on Cantor's part of the actual infinite, in opposition to a long tradition originating with Aristotle that, with a few exceptions – one of the most important being Bernhard Bolzano's *Paradoxes of the Infinite* in 1851 – had assigned to the infinite a merely potential significance. In the *Fundamentals*, Cantor made a distinction between *reelen Zahlen*, real numbers that are distinguished from complex ones (such as $\sqrt{-1}$) in a mathematical-formal sense, and the *realen Zahlen* – numbers that have a real existence rather than a purely formal one, even when introduced in a purely formal way. The transfinite numbers, just like the irrational ones defined by the fundamental sequences, consequently had to possess a wholly analogous ontological status; the irrational in relation above all with points in geometrical space, the transfinite in relation to a material reality consisting of a combination of monads of (still) uncertain cardinality. The transfinite numbers were to be considered, Cantor noted, as new irrational numbers. Their mathematical definition, in formal terms, placed them on the same, identical level. Irrational and transfinite numbers resembled each other in their intrinsic nature, both were forms or variations of the actual infinite, and if we did not accept the one then we were forced, for consistency's sake, not to accept the other.[6] Defining the numerical atoms of the continuous line as fundamental sequences, of an infinite kind, of relations between whole numbers, Cantor thus inherited the atomistic realism of the Pythagoreans. The *arithmós*, the positive whole number, now seemed capable of founding a mathematics of the indivisibles, of the number-points of the continuum.

Gottlob Frege defended Cantor's thesis, comparing the

ontological status of numbers of various kinds in the name of logical coherence:

> I find myself completely in agreement with him in judging the opinion of those who, among all possible numbers, would like to recognize only finite natural numbers as actual . . . In our research we may use, without the least reservation, whatever name or sign that has been introduced with a logically impeccable method; so that the number ∞_1 ends up being as justified as the numbers 1 and 2.[7]

For Frege, all numbers that were introduced in a logically coherent way were *actual.* This was a guarantee of reality that the rational and irrational numbers now provided for the continuum and for all the theories of analysis that made it possible to establish that certain mathematical entities, such as the point of intersection between two lines, actually *exist* in the continuum. This was a position which overturned the whole theory of space-time maintained up to this date, and refuted the most widespread belief among philosophers, at least since Leibniz, that a space made up of points is not logically possible (Russell, *Principles*, par. 423). A crucial moment in the development of this theory had been Kant's *Critique of Pure Reason*, in particular the second antinomy (or coexistence of two contradictory assertions that may both be justified) discussed therein, on the composition of substances. The thesis of the antinomy consisted of the affirmation that 'every substance that is composed consists of simple parts, and nowhere does there exist anything other than the simple, and that which is composed of it.'[8] Russell was able to explain that, thanks to recent arithmetic theories of the continuum, the contrary, antithetical affirmation was no longer sustainable. The argument used against the thesis and against the existence of points, he noted (*Principles*, par. 435), would show that neither should numbers exist, rational and irrational, that in Cantor and Dedekind were associated with points on a straight line. In the *Doctrine of Science* (315, 7) of 1837, Bernhard Bolzano had already confronted the question, arguing that the demonstration of the Kantian antithesis, according to which no composite substance consists of simple parts, was completely mistaken. Bolzano conceived of the continuum as an aggregate of simple and indivisible entities, such as points or instants. Taken as a whole, his theory was inadequate: in the absence of a criterion of *completeness*, the lacunae of the rational corpus remained. Nevertheless, Bolzano anticipated – including in his *Paradoxes of the Infinite*, published posthumously in 1951 – the numerical atomism that would go on to characterize the more rigorous theories of the arithmetical continuum that appeared towards the end of the century,

Henri Bergson was aware of the demands of a realism that was both practical and philosophical that was implicit in punctiform components of space and time. Bergson had been struck by the realization that lived time, a matter so central to philosophy, had thus far been completely ignored by mathematics.[9] Time and space, for mathematicians, were placed on the same level, as if they were things of the same kind, and all that was needed to pass from one to the other was to exchange 'succession' for 'juxtaposition'. Our intelligence, Bergson noted, operates with more ease when it has to deal with points that are in some way fixed: it asks itself where a moving thing finds itself in a certain instant, where it will be in a subsequent instant, and through where it will pass; yet even if it appears to be so interested in temporal duration 'it always wants to deal with immobility, real or potential'.[10] But with Bergson the meaning and the scope of the mathematical continuum are overturned: precisely due to the fact of being real or actual, in order in the end to respond to our practical needs, mathematical time and space betray lived experience and interior time. It was then Norbert Wiener, during the course of his collaboration with Eberhard Hopf, who realized that Bergsonian time – not a linear flow but an '*emboîtement* or nesting of mental events one within another, in the gradual enrichment of the I'[11] – served also to explain the physics of stars and the functioning of the atomic bomb. Together with Hopf, Wiener introduced a class of integral equations (known as Wiener–Hopf equations) that would go on to be used to simulate temporal processes of the most varied kind, with efficient techniques of signal prediction and filtering. Also in the background of Wiener's activities and research, with the study of devices and mathematical models based on the concept of feedback, an orientation emerged that was in keeping with the vision of Proust and of Bergson. The time of cybernetics, which differed in some ways from that of physicists, had also to be that of teleological phenomena, of processes of growth and learning – and so resembled duration as it was really experienced, as interior time organized as a process of reciprocal (inter)penetration of states of consciousness. Its structure, based on the phenomenon of *emboîtement*, could be expressed in the mathematical models of temporal processes,[12] in terms of integrals and Toeplitz matrices, which expressed the same idea of the growth of figures – by way of successive gnomonic corrections – as in the mathematics of antiquity.

$$\begin{bmatrix} a & b & c & d & e & f \\ g & a & b & c & d & e \\ h & g & a & b & c & d \\ i & h & g & a & b & c \\ l & i & h & g & a & b \\ m & l & i & h & g & a \end{bmatrix}$$

The form of Toeplitz matrices (here written for only six rows and six columns) is characterized by the equality of all the elements on the lines that are parallel to the principal diagonal, hence the first row and the first column are sufficient to define it. The phenomenon of *emboîtement* is visible in the fact that inside the matrix it is possible to segment smaller matrices of exactly the same Toeplitz form. This property of self-similarity (of an object that is exactly or approximately similar to a part of itself) recalls similar constructions in Greek geometry, reflected in this case in the structure of the inverse matrix,[13] which can self-generate from the first row and column.

Toeplitz matrices, the direct consequence of the discretization (transforming continuous functions into discrete counterparts) of Wiener–Hopf equations, would acquire over the years a great importance in matrix algebra, presenting themselves as an archetypal point of reference for the algebraic concept of the *structure* of a matrix. In such a structure one had to look for the reasons for computational efficiency, but also the reasons for the affinity between lived time and mathematical time.

Consequently the arithmetical continuum of real numbers provided the foundation for the very same applied mathematics, and the numbers, ordered forms corresponding to precise models of physical or biological reality, could recall by analogy the reality of lived time, of the authentic subjective time that Bergson had contrasted to the spatialized time of the mathematicians. This depended, Wiener believed, on single and specific processes that can assume at will a physical, chemical, physiological or mathematical significance.

Notwithstanding this, it was soon discovered, thanks above all to Brouwer, that the fundamental series that served to define real numbers, in the new theory of the arithmetical continuum, do not respond to a criterion of computational effectiveness. This proved to be a crucial finding, tied to the crisis in the fundamentals of mathematics, in the course of which actuality and effectiveness would become the prerogative of another concept that the mathematicians of the twentieth century would assume the task of defining – the concept of algorithm.

12. The Reality of Numbers: Dedekind's Sections

The importance of logic in the development of mathematics is universally acknowledged. But can it be said that mathematics *derives* from logic? The question is different from the one regarding the *reducibility* of mathematics to logic, which between the end of the nineteenth and the first part of the twentieth centuries Richard Dedekind, Gottlob Frege and Bertrand Russell had sought to answer in the affirmative.

If we look at the principles that in antiquity founded the disciplines of both mathematics and logic, we remain caught in a dilemma: does the reasoning on which the syllogism is based have its own independent origin and justification, or is it born out of mathematics instead? Relatively recent investigations have demonstrated at least the plausibility of the second hypothesis: a careful interpretation of the syllogistic theory of Aristotle (*Prior Analytics*, 25 b 26 ff.) already makes it possible to recognize a kind of reasoning and a corresponding selection of terms that take us back to the Greek theory of proportions.[1]

What can be said, in turn, about the theory of proportions? Are we dealing with a theory of mathematics, or with a theory of logic, thought through with, and dressed up in, mathematical terms? The question is not a specious one, because the Euclidean theory of proportions could pass for logic in disguise (something implicitly contended already in 1928, by Hans Reichenbach)[2] – and because the definition of proportion in Book V of Euclid's *Elements* gave rise, at the end of the nineteenth century, to an extension of the apparently *logical* nature of the concept of number.

As is already evident in the example of lateral and diagonal numbers, a real number can consist of an infinite sequence of intervals with rational extremes, each one contained within the preceding one, the length of which tends progressively towards 0. On the basis of a proposition, one assumes that a point exists that is contained in all of the intervals. Besides, given the sequence of intervals, we can find a criterion for dividing the corpus of the rational numbers into two classes, such that every number of the first class is inferior to any number belonging to the second.[3]

Figure 7

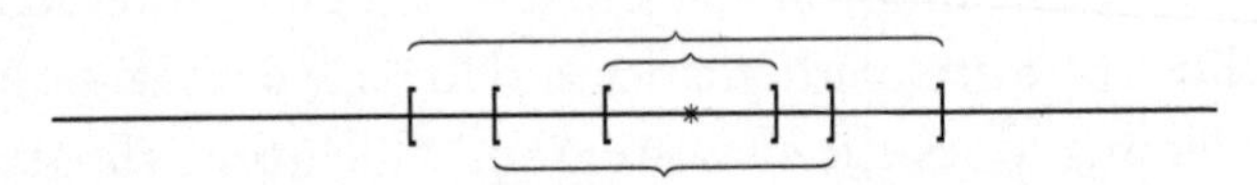

Based on this schema, models of the arithmetical continuum would be elaborated towards the end of the nineteenth century that were substantially equivalent to that of Cantor's fundamental sequences. Dedekind,[4] in particular, would explain to an exhaustive degree that an irrational number x is defined every time that we have a criterion for assigning every rational number to one and only one of two classes of rational numbers A and B, where every number of A is inferior to every number of B and, moreover, where A does not possess a maximum element and B does not possess a minimum one.[5] In the language of Dedekind, the pair (A, B) defines a *section* (*Abschnitt*) of the corpus of rational numbers. Now the theory of proportions of Eudoxos/Euclid has the same meaning. It establishes that, given four quantities a, b, c and d, the two relations $a{:}b$ and $c{:}d$ are equal if $a{:}b$ is greater than, equal to or less than the relation $m:n$ (m and n whole numbers) every time that $c:d$ is, respectively, larger than, equal to or less than $m:n$.[6] In such a case the relation $a:b$ divides the set of the rational numbers into two classes A and B, the first formed by all the relations $m:n$ that are inferior to $a:b$ and the second by all the relations $m:n$ superior to $a:b$. In a similar way it is possible to define two classes C and D in correspondence to the relation $c:d$, and we can then demonstrate that $A = C$ and $B = D$. In other words, if the relations $a:b$ and $c:d$ are equal in Euclid's sense (*Elements*, V, Def. 5), then they identify the same section, that is to say, the relative pairs of classes that define the sections are co-extensive.

The idea of the section is relatively abstract and lends itself to a logical definition of real numbers such as that proposed by Russell or Quine; but we have seen how a typical example of a section is derived from the algorithm that calculates the relations $d:l$ between diagonal and lateral numbers to approximate $\sqrt{2}$. Of course, this algorithm was one of the many possible ones, conceived of in remote eras, which might have given rise to the idea of Eudoxos/Euclid, and after them to that of Dedekind. But it has a paradigmatic aspect, because implicit within it are the principal concepts from which a definition of real numbers was extrapolated. The property of *completeness* of real numbers that expresses the absence of gaps can be formulated by asserting that *every Dedekind section is generated by a real number.* The property of completeness may be defined in many equivalent ways, and can be attributed to any domain K such that every one of its elements generates a section in K, though it is evident that its reason for being is founded on the existence of algorithms which can approximate irrational numbers. The *completeness* (every element generates a section) and the *density* (given in any case two distinct elements a and b, where $a < b$, there exists another element greater than a and less than b) are the two essential properties for an ordered set to be said to be continuous. The rational numbers form a dense but not complete set; the whole numbers form a set which is not dense but is complete; the real numbers form a set that is both dense and complete.

But we are prompted to ask ourselves: what need is there to define $\sqrt{2}$ as a section of the rational corpus rather than as a number (of infinite decimal digits) that when squared gives 2 as a result? We do not exclude the possibility of expressing an irrational number as an (aperiodic) sequence of infinite digits, nor the possibility of defining an irrational number, in some specific cases, as the number that raised to a certain power gives a whole number as a result. But behind the concept of *section* there is above all the idea of the *relation* between two quantities ($\sqrt{2}$ and π are the relations, respectively, between the diagonal and the side of a square, and between a circumference and its diameter – and Dedekind resorts explicitly to the Eudoxos/Euclid theory of relations), and to say that this relation corresponds to an irrational number means that the latter is not expressible as a relation between whole numbers (*Elements*, X, 7).

Hence, in order to understand what an irrational number is we must first know what a relation is, that is to say, what meaning can be attributed to the relation between two quantities that is expressed with the sign ':'. To define the concept of relation we may rely then on two theories: that of Eudoxos/Euclid, which originated in *c.* 350 BC (*Elements*, V, Def. 5) and to which Dedekind refers, and the one to which

Aristotle alludes, referring to the geometrical idea of likeness (*Topics*, 158 b 29 ff.):

> In mathematics, too, some things would seem to be not easily proved for want of a definition, e.g. that the straight line, parallel to the side, which cuts a parallelogram divides similarly both the line and the area. But, once this definition is stated, the said property is immediately manifest; for the (operation of) reciprocal subtraction applicable to both the areas and the lines is the same (or gives the same result); and this is the definition of the 'same ratio'.[7]

Now, as we have seen in the case of the diagonal and the side of the square, the *anatanaíresis* consists of a process of elementary operations of measurement that resolves itself every time in a subtraction of one quantity from another. To measure a magnitude x with another y that is smaller, taken as a unit of measurement, means in simple terms, as Bernhard Riemann noted in 1854,[8] *superimposing* one quantity over another and fixing the number q of times that the smaller quantity y goes into the larger x (q for the quotient). If there is a remainder r other than 0 (smaller than y), the operation of measurement of y and r is repeated, which is to say that we count how many times r is contained within y. The process is repeated indefinitely in the case of incommensurable quantities, because no remainder is nil. This means, in effect, *dividing* the magnitude x by the magnitude y. With the quotients q obtained in the course of the process we construct, as will later be clarified, the *continuous fraction* that *defines* the irrational number corresponding to the relation between the incommensurable quantities x and y.

The critical phenomenon that involves the concept of section is now the following: with the *antanaíresis* we construct, step by step, with the numbers q that establish how many times a quantity is contained within another, numerical fractions $m{:}n$, the value of which is alternately greater than or less than the relation $x{:}y$.[9] In other words, the *antanaíresis* serves to define two successions A and B of numerical fractions, increasing and decreasing respectively, where every fraction in A is smaller than the relation between x and y, and every fraction in B is larger. The fractions in A are approximations by defect, those in B are approximations by excess of the same ratio $x{:}y$.[10] We could then decipher the Euclidean definition of the equality of relations by giving to the numerical ratios $m{:}n$ involved in the definition the same meaning as the numerical ratios constructed with *antanaíresis*. The equality of the two ratios between *magnitudes* is decided on the basis of the order relation ($>$, $<$ or $=$) that such ratios have with the ratios between numbers; but if we want to assign a real existence to these ratios we need to imagine them as the result of a process of calculation by means of *antanaíresis* or of another procedure with a sufficient degree of efficiency.

When Simone Weil suggests in a letter to her brother André that the discovery of incommensurability was not only traumatic but also a cause for celebration, we need to think of the calculability of the numerical relations: it is possible 'to see that what is not *defined* through numbers is nevertheless still always a relation',[11] thanks to the fact that the absence of a definition through numbers is in any case offset by procedures that calculate numerical approximations by excess and defect.

The abyss of the infinity of the continuum is certainly not exhausted by the procedures that approximate magnitudes with numerical ratios – it isn't, that is to say, exhausted by the set of numbers that may be computed, which, as Turing demonstrated in 1936, is only numerable;[12] but the entirety of those procedures permits an approximation to the abyss of the *ápeiron* that does not entail an irreparable loss of limit and unity. As Plotinus stated (*Enneads*, VI, 6, 3), 'if you rely on the infinite without throwing over it, like a net, some kind of delimitation, it will escape from your grasp and you will find nothing that is unitary, because if that were the case you would already have defined it'. In the Orphic and Pythagorean tradition, from the atomists and Plato up to Nicomachus at least, whole numbers and their relations were the constitutive elements, the real knots of this net.

When Russell and Whitehead in *Principia Mathematica* (1910–13) tried to define mathematics in terms of logic, and in particular an irrational number in terms of classes, they were thinking above all of Dedekind's concept of the section. Later on, Quine did more or less the same,[13] taking as a model the definition elaborated in the *Principia* – a sure sign that the theories of Dedekind and Euclid lent themselves particularly well to the formulation of a theory of real numbers in purely logical terms. Russell explained this at length in the *Principles of Mathematics* (pars. 259 and 261). Russell's vision presupposes a theory of universal and existential quantifiers, based on the use of terms such as *all a*, *some a*, *every a*, *an a exists*, that reconnects the knowledge of real numbers to a logic of propositional functions, and definitively to a system of assertions that may be formally deduced from a collection of axioms. Furthermore, the logic of the quantifiers is validated by the *mathematical* idea of the section and by the algorithmic schema that in turn preceded and justified it (par. 60).[14]

To what extent does the concept of section correspond to

our intuitive idea of numbers? Dedekind's theory is in some respects counter-intuitive, because it does not seem natural to conceive of a number as a collection characterized by a relatively complex structure such as the section has. For this reason, in case the section did not identify a rational number, Dedekind considered it necessary to *create* a new entity, an irrational number α, where α is a symbol designed to denote the pair of classes A and B. Hence Dedekind's creations have an affinity with Cantor's. He noted as much in a letter of 1888 addressed to Heinrich Weber:

> *It is exactly the same question, with regard to which you claim that the irrational number should be none other than the section itself, while I prefer that something* new *is created (different from the section) that corresponds to the section and* produces *the section. We have the right to attribute this power of creation to ourselves, and it is much more appropriate to proceed in this way to tackle all numbers. Even rational numbers* produce *sections, but I will of course not seek to identify rational numbers with the sections that these* produce.[15]

The emphasis falls here on *producing* (*hervorbringen*), a capacity attributed to the intrinsic properties of numbers. Numbers *generate* the sections thanks to a kind of potency, a *dýnamis* that makes one think again of the theses of Plato's *Sophist* (247d) and which thereby qualifies them as actual and efficacious entities. The reality of numbers arises from this, from their ability to impose new lines of thought that prefigure new concepts in turn. The freedom to create concepts seems to adjust to a strict necessity, and reality manifests itself every time we encounter such a necessity.

Dedekind's theory answered the criterion of *completeness*: it seemed to appeal, with proof, to our intuitive concept of continuity, while still not corresponding to a criterion of graduality. Together with *density*, as previously mentioned, completeness characterizes the numerical continuum of real things and is a technical synonym for the absence of gaps: with the exclusion from the domain of numbers of sections that represent irrational numbers, there would be introduced gaps the presence of which cannot be allowed in a continuous set.[16]

For Cantor also, the sign that denotes the fundamental sequence was introduced by means of an act of creation – not of an arbitrary kind but due to the *necessity*[17] for an extension of the corpus of rational numbers. The notion of a number of elements of a set has an immediate objective representation, Cantor explained, and 'the relation between the number of elements of a set and a number itself demonstrates the *reality* of the number even when it is infinite'.[18] With their theory of the numerical continuum Cantor and Dedekind overturned the Kantian thesis of the supposed a prioristic nature of the perception of time and space for mathematics: the concept of the continuum was more primary than either space or time and was basically founded on the idea of number and class.[19] A further circumstance gave credibility to the *reality* of the new numbers. If we forgo a requisite clarification of the so-called Archimedean property of a numerical field[20] (i.e. of having no infinitely large or infinitely small elements), Dedekind's theory of the continuum becomes equivalent to Cantor's. To assert that a numerical field K is Archimedean, and such that every fundamental sequence has a limit in K, is equivalent to asserting that in K every Dedekind section is generated by an element s belonging to K.[21]

In the last decades of the nineteenth century, everything seems to converge on the assertion that numbers are real entities. During an exchange of letters with Cantor, Charles Hermite confessed that whole numbers seemed to form a world of realities that exists outside ourselves, with the same character of necessity as natural realities that offer themselves up to be apprehended by our senses. For Cantor the reality of numbers – whole, rational, irrational and transfinite numbers – was founded, as well as on concrete evidence, upon the same truths as the Scriptures. The latter seemed to testify to a divine knowledge of infinite numbers: Augustine explained it in *City of God* (XII, 19), which for Cantor became a support in favour of his mathematical theory of the actuality of the infinite. The correspondence between numbers and knowledge, proclaimed by Augustine, was legible by litotes in Psalm 146, 4–5, in the passage that is sung of Him who 'calculates the number of the stars / and gives to each a name' but who is also the Lord 'whose intelligence is without number [*tês sunéseos autoû ouk ésti arithmós*]'. For the Lord of the universe, according to Augustine, the infinity of numbers is not incomprehensible, 'even though there is no number for the infinite number' (*City of God*, XII, 18).

The same idea of the section, at the end of the nineteenth century, was also circulating among Italian mathematicians, and Ulisse Dini in particular made it the foundation of his theory of real functions. He set himself the task of elaborating an arithmetical theory of the continuum, exactly as Cantor and Dedekind had, that was independent of geometry and any presupposition based on perceptions of space and time. His theory is in every way similar to that of sections, and in fact Dini and Dedekind cited each other. For Dini also, the two classes that identify an irrational number, one of decreasing and the other of increasing order, are constituted by numbers that approach each other indefinitely and at a unique and determined magnitude the *existence* of which is known *a priori*, but which may never be reached with only consideration of quantities corresponding to rational numbers of the two classes, and which marks the limit between quantities of these classes; hence if to this quantity a corresponding number is also assigned, then this number will not be rational but something will correspond to it . . . effectively something real, given that a quantity corresponds to it that has a real and proper existence and that, although it may never be reached by way of the single quantities corresponding to rational numbers, is such that it may nevertheless approximate it as required.[22]

Dini repeatedly alludes to the real existence of numbers to indicate that they are not mere fictions, and that the *symbols* by which they are denoted, for the sole purpose of respecting a 'simplicity of locution',[23] correspond to an actual and tangible phenomenon, even if, we might add, one of unknown origin.

And yet something peculiar occurred. There was such certainty about the coherence and real existence of numbers that it seemed plausible to extract from them the general principles of a mathematical logic of finite and infinite sets. Towards the end of the nineteenth century Dedekind's theory of numbers had precisely the aim of evidencing a logical theory of sets as a prerequisite of the intuitive notion of numbers and of the recursive structure of arithmetical operations. But this logic dared to depart from the consented limits. There it came up against the wall of paradoxes, and mathematicians began to call into question the existence of classes. Realism gave rise to a sort of sceptical nominalism that relied on the coherence of the language of logic with essentially mathematical properties that had been known for centuries and perfected according to modern theories of the continuum.

The aim was to seek to derive everything from logical principles, but it is not likely that the reality of numbers depends on a logic of quantification that uses expressions of the type 'x exists such that $f(x) = 0$', or 'every x satisfies a certain property P' – the kind Russell used when dealing with Dedekind's sections. On the contrary, the reduction of mathematical properties to the logic of propositional functions has imposed a nominalistic vision which divested numbers of reality, shifting every one of their evidentiary features on to a matter of accuracy of linguistic formulas and the quality of coherence of a formal system.

Russell elaborated a *theory of descriptions* with the aim of making grammatical subjects literally disappear, reducing them to 'incomplete symbols', symbols that do not represent anything at all. It thus became possible to exorcize the problem of the existence of the 'hippogriff', or the 'squared circle', that otherwise risked being accorded the status of plausible objects in common language usage. Classes, and numbers defined by classes, do not escape either from this kind of annihilation. The symbols that denote classes, like those that are used in descriptions, are incomplete symbols. What matters is the correctness with which they are used, but in and of themselves they mean nothing: 'hence classes, in the way we introduce them, are mere linguistic or symbolic conveniences, not authentic objects . . .'[24] In a treatise on the logic of sets, Quine declared that his task consisted above all in deciding what the propositions are that identify a class or, in terms of a philosophical realism, 'which classes exist'.[25] But in the long run every kind of realism seems hazardous.

In reality, for Quine, the classes do not exist at all, logical formalism must always have as its objective the elimination of every explicit usage that implies their existence as real entities. Consequently, any real form of knowledge ends up being based merely on empirical observation.

What caused this diffidence with regard to the existence of numbers? Certainly the crisis in the foundations of mathematics – which came to a head at the beginning of the twentieth century in the wake of the discovery of paradoxes – intensified the suspicion already expressed by Leopold Kronecker as to the existence of infinite numbers. Cantor had abused the mathematician's power of abstraction: this was also the opinion of Frege, according to whom there was a risk that abstraction would be used as almost a magical power, and that transfinite numbers constituted entities made to appear by inadmissible conjuring tricks.[26] The discovery of antinomies vindicated the judgement of those mathematicians who had attributed to Cantor a groundless aspiration to omnipotence.

It is nevertheless striking to witness the brusque transition between the assertions of mathematicians such as Dedekind, Cantor and Dini, and the nominalism that was ushered in by logicism itself. The latter, we might say, assumed the responsibility of rendering harmless the embarrassing question that had been posed for two centuries, and for which Nietzsche had found the most radical formulation:

> the question remains open: are the axioms of logic adapted to the real, or are they criteria and means for *creating* the real, the concept of 'reality' for us? . . . In order to assert the former it is necessary, however, as has already been said, to know already what being is – and this is absolutely not the

> case. The principle therefore does not contain a *criterion of truth* but merely an *imperative* about that which MUST *be taken* to be true.[27]

Moreover, in keeping with a long tradition stretching from Aquinas to Descartes and William James, and as Whitehead and Russell made clear in the introduction to *Principia Mathematica*, definitions have a *willed* quality imposed by the choice of that which appears most worthy of being known. And yet behind this act of volition there was an imperative even stronger than the one intended by Nietzsche: the evidence of those unalterable and unmodifiable phenomena, connatural with the same numbers and with the same mathematical formulas, that are imposed by the most established and meaningful theoretical choices.

13. Mathematics: A Discovery or an Invention?

Whether mathematics is a discovery or an invention is a question often posed and abused, seeming always to require a critical response, a conclusive distinction between that which is real and that which is subjective and arbitrary. Regarding the possibility of establishing such a distinction, both mathematics and logic are destined to be called to bear witness. In fixing axioms, in demonstrating theorems or in constructing the solution to an equation, thought seems to be adhering to the real, and the will to necessity: an exchange in which an attempt is made, due to their immediate and originary character, to distinguish two terms that are antithetical at the precise instant at which they begin to interact. But their initial combinatory activity is already an inextricable tangle, and the question concerning what the *prius*, or forerunner, really is, between will and necessity, remains unanswered.

The need to introduce real numbers seems to be antithetical to the thesis, advanced by Dedekind, that it is a matter of 'free creations of the human mind',[1] but this freedom must surely correspond to a norm in keeping with those that Dedekind called 'the laws of thought', the congenital ability of our mind to connect one thing with another, to make one thing correspond to another, to represent one thing with another; an ability without which it would be impossible to think. From the moment of our birth we exercise these connective and representational operations continuously, but *without a predetermined objective.* The chains of reasoning seem as a result to arrange themselves, naturally, so as to suggest apparently simple notions, which are in fact complex, such as the abstract notion of number, which articulates itself through successive generalizations until it reaches real and complex numbers. Every algebraic and analytical theorem, Dedekind noted, however advanced, may be expressed with a theorem on whole numbers, but this circumstance must not impede the creation of new entities:

> I can see no merit – and Dirichlet thought this too – in really fulfilling this wearisome circumlocution, and in using and recognizing with insistence nothing but rational numbers. On the contrary, the greatest and most advantageous progress in mathematics and in other sciences has invariably consisted of the creation and introduction of new concepts, rendered *necessary* by the frequent recurrence of complex phenomena that may only be controlled with difficulty by the old notions.[2]

Poincaré used not dissimilar arguments to justify an arithmetical definition of the continuum. Hence, the reason that pushes us to new creations does not depend on volition alone, but on the frequency of occurrence of objectively complex phenomena which it is convenient to label by way of new concepts. But the implicit reality in that which allows itself to be designated by these concepts reveals itself above all in the *capacity to produce them.* In this way we go back to a sentence of the Platonic *Sophist* (247 d–e), according to which 'entities are nothing other than the power to produce', and to the mathematics of the *Śulvasūtra*, the Vedic treatises that posed, among the principal problems, the production of sequences of figures of increasing scale. This idea recurs in the work of Dedekind on various occasions: the number produces the section in the arithmetical context in the same way in which a corresponding point produces it on the geometrical straight line.[3]

We find the same thesis on the existence of the number that generates the section of the corpus of rational numbers in the mathematics of Karl Weierstrass. In the annotations of Weierstrass's lectures collected by Salvatore Pincherle, it is explained among other things in what way we may establish a connection between numbers and points on a geometrical straight line. If you fix the origin O, to link a number to the segment of the line OB, you measure OB as a function of the preassigned unit OA, by means of a procedure that is analogous to the *antanaíresis* or Euclidean algorithm that permits the calculation of a number a such that $OB = aOA$, and a, the number associated with OB, can be rational or irrational. Vice versa, to associate to every finite and determined number a a segment of the line OB measured by a, one will have to conceive B as the point of separation between two classes of points, corresponding to numbers that are greater and less, respectively, than a. An idea analogous to Dedekind's, though not explicitly oriented towards the definition of a real number as a section. In the specification of the concept of equality between two numbers (rational or irrational) Weierstrass reclaims a form of reasoning equivalent to that which is required to define the equality of two relations in Eudoxos/Euclid's theory (*Elements*, V, Def. 5), which Dedekind in turn would resort to in order to define the concept of numerical continuity.[4]

Decisive in defining the arithmetical continuum was the idea of denoting the numbers that produce sections with symbols. With the help of symbols, Whitehead noted, we are capable of carrying out transactions that are almost mechanical in the course of a reasoning that would otherwise require the most complex neural procedures to be undertaken. Nor should one cultivate the habit of thinking about what one is doing every time one does it. Actually, the reverse may be advisable, since advances often happen by 'extending the number of important operations that we can conclude without thinking about it'.[5]

In his celebrated essay of 1940 entitled *Apology of a Mathematician*, Godfrey H. Hardy confronted the theme of the reality of mathematics without too many hesitations and in a manner that he himself defined as dogmatic, the aim of which was to avoid any misconception. He claimed to believe, in accordance with the majority of mathematicians, in a *mathematical reality* different from the physical one, but observed that there was no agreement as to its nature: some argue that it is mental, that we are the ones who construct it; others are convinced that it entails an external world that is independent of us. He pointed out that an acceptable definition of mathematical reality would certainly help to resolve the problems of metaphysics, and if an explanation of physical reality were also included, why then, all problems would be resolved. Hardy expressed his convictions as if they were set in stone:

> I believe that mathematical reality exists outside of us, that our task is to discover or to *observe* it, and that the theorems that we demonstrate, qualifying them pompously as our 'creations', are simply annotations to our observations. This opinion, one way or another, has been supported by many great philosophers, from Plato onwards, and I use the language that is natural to one who shares it.[6]

Countless proofs and testimonies attest that, no sooner have they presented themselves as evidence, than the entities created by a mathematician open up an unknown territory where every *discovery* is far from capable of being considered, along with those entities, as the product of free will. But it seems to reinforce the fact that neither is the first act of creation free from an intrinsic necessity that obliges it to be taken as more of a discovery than an invention. The definitions of real numbers proposed at the end of the nineteenth century are part of a single chain of ideas the reasons behind which are anchored in a set of properties pertaining to numbers and their relations already observed in ancient times, specifically designed to impose on numbers those particular definitions and not others. In these properties the theme of growth is included, and the study of the enlargement of figures that so preoccupied the mathematicians of antiquity allows us to glimpse what would go on to become the central subject of modern computation: the growth of numbers as a critical phenomenon for the stability of digital calculation.

14. From the Continuum to the Digital

Dedekind explained that if space has a real existence, it does not necessarily follow that it should be continuous, because many of its properties would remain the same even if it were discontinuous.[1] The mathematical reality of space was consequently independent of the existence of irrational numbers: their creation as sections of the rational corpus embedded itself in a reality already formed and acknowledged, which was evidently founded on whole numbers and on their relations, the true antidote to the non-being of the infinite, to the absence of structure inherent in the *ápeiron*.

In Dedekind's words, one gleans a kind of justification *ante litteram* of the theses upheld by the most intransigent constructivist mathematicians, such as Leopold Kronecker and Henri Lebesgue, for whom the analysis could well do without the sets of numbers, such as the field of real numbers, of superior power to the numerable. This was a premise that allowed one to think of the continuum not so much as a completion of the discrete as an extrapolation of it, essentially unreal, imposed by a project of simulation of that which appears intuitively, in our everyday experience, as a space and a matter that are full, without jumps and without gaps. Was it not perhaps affected, that project, by our insuppressible demand for *graduality*, that is to say, for a continuum conceived as 'free becoming'[2] rather than as an assemblage of completed entities?

Like Dedekind, Kronecker conceived of irrational numbers as the product of a division of the rational corpus into two classes – that is to say, as a section, but a number such as $\sqrt{2}$ was nothing other than a sign activated to denote this division. Henri Poincaré in *Science and Hypothesis* (1902) asked himself if in order to be sufficiently satisfied with a mathematical theory of the continuum, we should *forget* the origin of these signs. But could mathematical knowledge really base itself on such a manoeuvre?

We are used to thinking of the discrete as an approximation of the continuum, in the sense that irrational numbers approximate sequences of rational numbers. But why not reverse the terms and think of the continuum, on the contrary, as an approximation of the discrete? What we really know are whole numbers and their relations, which we call rational numbers – our knowledge of reality is based on these numbers, all calculations are rational, and perhaps there is no need to extend our knowledge to numbers that, while filling the gaps in the rational corpus, could turn out not to be essential in the actual operations of a human or digital calculator.

Of the atomism implicit in the arithmetical continuum we find a radical criticism in late Nietzsche, albeit shifted into the physical realm:

> Physicists believe, in their way, in a 'real world': an atomic system which is fixed for all beings, with necessary dynamics – so for them the 'world of appearances' is reduced to this aspect, accessible to every being in its own way, of universal being and universal necessity . . . But in this they are deceived: the atom that they postulate is derived from the logic of perspectivism of consciousness, and is consequently itself a subjective fiction. This image of the world that they draw is in no way essentially different from the subjective image of the world: it is just constructed with more

> developed senses, but they are always *our* senses . . . And in the end they have neglected something in the constellation without realizing it: in effect the necessary perspectivism by virtue of which every centre of power – and not only man – constructs the rest of the world on the basis of itself, that is to say, measures, models and shapes it according to its lights . . . They have forgotten to factor into 'real being' this force that *creates* perspectives.[3]

The statement of principle launched by Nietzsche was a sign of the times and captured a typical motif of modernity. It was certainly not seeking to undermine the aim of elaborating a mathematical model of the intuitive continuum, of that kind of assemblage that Poincaré discerned with good reason in the theories of Cantor and Dedekind. But at least it contributed to discrediting those theories, cutting down to size the pretence of establishing that something called section, irrational number, atom or point exists *in reality*. That which Cantor, Dedekind and Poincaré retained as almost a constraint – the peremptory need to add to sentience (and to its amplifiers as furnished by instruments of ever greater accuracy) the direct knowledge of a 'real' albeit invisible world – could now change into a kind of scission or double perception, in a problematic tension between a more radical claim to actuality and an abstraction that reaffirmed subjective fictionality.

Hilbert believed that mathematical abstractions and the fictions with which we think about the infinite (including the arithmetical continuum) must have a foundation in reality by way of a 'pre-established harmony'.[4] But a more compelling criticism now emerged from mathematics itself: the paradoxes of the theory of sets, starting with the famous antinomy that Russell communicated to Frege in 1902, made questionable the indiscriminate existence of classes and activated, among mathematicians, the search for new foundations and new grounds for certainty.

A few clarifications emerged from the relative ease with which it was possible to unmask the causes of certain paradoxes. The celebrated Richard's paradox, in particular, was the consequence of a cognitive act that to all intents and purposes was legitimate: to aggregate in a single class all the numbers that may be described, in a language, with a finite number of words. But such a class does not exist; it is unreal, as Émile Borel would remark in 1908, because it is not enough to assign to a set of words the attribute 'finite' in order to ensure the actual existence of the entity that those words want to describe. *Finite* had to be, properly speaking, a process of calculation that from a set of data was capable of elaborating a result in a limited number of stages. Borel noted that the existence of a mathematical entity can only be guaranteed by its actual construction. The real roots of an algebraic equation, for instance, may be calculated using a procedure capable of evaluating the decimal sums up to the degree of accuracy required. That which guaranteed its existence, therefore, was an *algorithm*.[5] Not by chance, Borel was considered the pioneer of the systematic study of algorithms that would have built a science of calculation based precisely on the concept of the algorithm and on the systematic use of numerical and information procedures.[6] In the same year of 1908, Zermelo published a study of the proper ordering of sets that was meant to ensure that they were constructible free from paradoxes. Not long after this John von Neumann elaborated an equivalent theory.[7]

In the first years of the twentieth century a critique of the theories of the numerical continuum emerged through mathematical intuitionism, with arguments that did not depend directly, at the time, on the urgent need to find a solution to paradoxes. The fundamental criterion of intuitionism was to acknowledge as extant and real only those entities that could actually be constructed. The arithmetic of rational numbers was consequently protected from intuitionist criticism, because the theories of Peano and Dedekind had given sufficient justification for the existence of natural whole numbers, guaranteed thanks to their effective construction by means of successive additions of units. Moreover, Dedekind, with the theory of recursion, or of definition by induction, had demonstrated that to the ordinary operations of mathematics a coherent and actual existence could be ascribed, and that the concept of calculability is based on that of enumeration, that is to say, on the series of natural numbers. The *existence* of an operation such as addition could be summarized in the idea that it is possible to actually add, with a series of finite steps, any two natural numbers – for instance $3 + 5 = 8$ or $346 + 512 = 858$; in general, $n + m$ for every pair of natural whole numbers n and m. Fundamental to this was the idea of *closure*, in the light of which one could discern a kind of control to the *growth* of numbers and quantities, already posited by the mathematics of antiquity. In passing from one number to the successive one with the addition of a unit, or in the addition or multiplication of two natural numbers, no matter how large, one always remained in the ambit of the same domain, that of natural numbers, which for this reason was called *closed*. One needs, however, to signal a critical passage: even given that the property of closure served to remove the danger of the growth of numbers, Dedekind was still oblivious to the problem of the instability of calculation, which would turn out to be the most insidious consequence of that growth. Soon it would no longer be enough to assure oneself that, while growing disproportionately, the numbers still belonged to the same domain.

The case of real numbers, however, at least at a theoretical level, was more complex than that of natural and rational numbers, because their definition implied the introduction of infinite classes of numbers together with the systematic use of existential and universal quantifiers, which is to say of propositions such as 'there is an x that satisfies the property P' or 'for every x the property P is valid'. And even if classical analysis, based on the concept of real numbers, did not seem to be compromised by the discovery of antinomies, it was still subject to criticisms and reservations on the part of those who, like the intuitionists, did not admit mathematical entities, numbers, sets or functions that were not actually calculable.

Once the rules of calculation had been established with rational numbers, real numbers were defined as fundamental sequences $a_1, a_2, \ldots a_n, \ldots$, of rational numbers that satisfied, we emphasize again, the following property: if one assigns a number ε, however small, *it is possible to find* a natural number k, dependent on ε, for which the distance between a_{n+p} and a_n, where $n > k$, is less than ε for every natural number p. But what does it mean, this expression *it is possible to find*? Intuitionist mathematics indicates only that *it is possible* to calculate effectively the index k for which the distance between a_{n+p} and a_n is less than ε if $n > k$, that is to say, that there exists a procedure capable of calculating that critical value of k.[8] This simple observation, to which Brouwer and the intuitionists resorted in order to attack classical analysis, was sufficient to render fictitious precisely that atomistic concept of number which Cantor had relied on to lend a contained quotient of reality to the mathematical continuum. To denote a fundamental sequence with a symbol was not enough to make the number corresponding to that symbol real, unless it were possible to calculate the index k by which the distance between a_{n+p} and a_n becomes arbitrarily small for $n > k$.

For the intuitionist mathematicians, effective calculus had theoretical bases that were far from the concreteness required by a science of algorithms orientated towards resolving problems of application in the most efficient way possible. Brouwer considered mathematics to be an action the reality of which was based on awareness and on the intuition of a multiplicity of sensations that developed over time. The intuitionist construction of a real number, conceived like a gradual definition, in time, of increasingly small intervals that include it, did not even have the materiality of the calculus of Turing, Kleene or von Neumann. Computational mathematics, theoretical and applied, would satisfy the demand for construction of ever more exacting criteria of efficiency.

The need to render effectively small, by means of a procedure, the distances between approximations near to the limit did not only respond to a theoretical question of how an irrational number is to be defined. Implicit is the demand for a new knowledge of the objectives of mathematics. Mathematical science must be capable of going in two opposite directions: from the concrete to the abstract, but also from the abstract to the concrete – and it is at the very least uncertain which of the two directions responds best to its deepest and most secret inclination. In effect, at least four lines of research lead us to think again about the extreme attention that mathematicians have devoted over the centuries to the relation that their theories entertain with the real world, be it material or invisible: the doctrine of number-points of the Pythagoreans and of the ancient atomists, mathematical physics, the theory of real numbers and the subsequent science of calculus based on the concept of algorithm.

As Giuseppe Peano observed, the aim of mathematics substantially consists of the calculation of sums of numbers that are solutions to the equations that simulate the most diverse phenomena of nature: 'The aim of mathematics is to determine the numerical value of the unknowns that manifest themselves in practical problems. Newton, Euler, Lagrange, Cauchy, Gauss, and all the great mathematicians develop their marvelous theories up to the calculation of the required decimals.'[9] Now, mathematics can carry out this task only if it is oriented towards the concrete as well as the abstract. The differential and integral equations, and the problems of the minimum, in which the mathematical models of both natural and artificial phenomena are expressed, have their origins in physical situations, economic and informational, and the nature of supplementary conditions on the boundary values and the initial values of the unknown functions is always motivated by the *physical reality* one is seeking to simulate.

Mathematics does not limit itself, however, to finding equations that define these models. The requirements that allow it to elaborate a solution need to be satisfied, and it is relatively rare that this solution can be expressed with an analytical formula, just as it is also improbable that one should use an analytical formula, assuming it exists, to find the solution of an equation in terms of wholly digital information.

The solutions written in analytical form, when they exist, are usually extremely complicated. Moreover, an analytical formula must satisfy the requirement of stability: to small variations of data, small variations in the results must correspond; otherwise, the information communicated by the model risks being completely distorted. This condition is not always satisfied, and we owe to Jacques Hadamard an example of initial values the analytic solution of which does not depend in a continuous way on the data. It is enough for these to be affected by a minimal error to have no useful information pertaining to the solution of the problem. Problems of this type have been termed 'ill-posed'.[10]

The demand that the solution of an equation exists and is unique, and that the problem should be well posed, is in keeping with a classic conception of mathematical physics, dominated by the presupposition that a physical event evolves in a stable and determined way, once the initial surrounding conditions have been established.

> Laplace's vision of the possibility of calculating the entire future of the physical world on the basis of an exhaustive set of data on its current state is an extreme expression of this tendency. Nonetheless this rational ideal of mathematical determinism based on the idea of causality underwent a gradual erosion when confronted with physical reality. Non-linear phenomena, quantum theory and the advent of powerful numerical methods have shown that 'well-posed' questions are very far from being the only ones capable of reflecting real phenomena.[11]

But it is not enough to assure oneself that the analytical solution, assuming it exists, is continuous with respect to the data. In reality, mathematical modelling entails various stages: the formula of the model needs to be approximated, usually, by an arithmetical formula, and the latter needs to be translated, by way of numerical procedures, into a purely digital calculation. In the most frequent case in which one approximates the differential model with a purely arithmetical one, such as a system of linear algebraic equations, one needs to examine one's perception in relation to the errors of data of this simplified model, as well as the stability of the process that calculates its numerical solution. In the end, calculus encompasses an enormous accumulation of elementary operations and a multiplicity of subprograms that constitute a *hidden* elaboration to which we do not have access but on which to a large degree the credibility of a computational process depends. As Beresford Parlett observed in 1978:

> These days most well specified computations are *hidden*. This means that the human user sees neither the data nor the output. In a big calculation the data for a subtask (a Fourier transformation, perhaps) will be generated by some program and the results promptly used by another. This is characteristic of introverted numerical analysis . . . Algorithms for hidden computations need to be much more reliable than those for which results will be seen by a human eye. Execution time seems to be less crucial but both reliability and efficiency are wanted. To what extent, in each case, can we have both? That is an interesting question.[12]

To summarize in broad terms, since the time of Leibniz and Newton mathematics has crossed three crucial moments, the third of which is currently still ongoing and is assuming the contours of what we are accustomed to refer to as the *digital revolution*. After the first discovery, in the seventeenth century, of that grandiose analytical machine that adopted the name of infinitesimal analysis, a second phase was initiated at the end of the nineteenth century, known as the *arithmetization of analysis*, equally revolutionary in kind: a search for the fundamentals of analysis in the concepts of number, of the set and of passage to the limit. In number, whether rational, real or transfinite, one needed to look for the ultimate reality, the authentic actuality of the finite, as well as of the infinite, for which mathematics had never ceased to search. Here, too, were to be found the seeds of the third revolution: the movement from first principles by which it was possible to trace back every form of reasoning relating to the concept of number to the implementation of a variety of methods for calculating sets of numbers. Algorithms and the lists of calculated numbers would inherit the information of initial models and would translate it, under the auspices of a genuine reductionism, on a different, more elementary, descriptive level. The theoretical possibility of returning analysis to the issue of number and to operations of the passage to the limit was to be developed alongside an actual arithmetization.

The final stage was due to two concomitant reasons: the crisis in the fundamentals of mathematics, which raised doubts about the possibility of indiscriminately using the concept of sets, and the surprising development of applied science, already under way since the end of the nineteenth century, which had among its effects the emergence, from the mid-twentieth century onwards, of information technology and scientific computation. Algorithms became an object of study for information theory, but also, from the second half of the twentieth century, the theoretical and practical nucleus of a science of computation on a large scale that would translate the problems of mathematical physics into a grandiose system of digital elaboration.

The principle of *reality* was consequently moved, in this third phase, on to the idea of *algorithm*. It was frequently noted during the course of this transition that mathematics cannot limit itself to procedures of abstraction but must also follow the opposite path – from abstract concepts to concrete reality: to the *earthly*, so to speak. From this very ground, from concrete applications to physical reality, mathematics was to derive the strength to free itself from the impasse of a crisis at its foundations that would otherwise have weakened its theoretical potency, Antaeus-like. The giant Antaeus derived his strength from his contact with the Earth, and Hercules managed to defeat him by holding him up in the air and picking him up every time he fell. Antaeus thus gradually lost his strength and was defeated and killed – but had he retained contact with the Earth, his strength would surely have prevailed.

Andrej A. Markov Jr, who made important contributions to the formalization of the idea of algorithm, had specific recourse to a principle of concreteness: 'Abstractions are indispensable in mathematics, and yet they should not be pursued as ends in themselves, nor should they lead to a point at which they do not descend to "earth". We should always think of the transition from abstract thought to practice as a necessary moment in human understanding of objective reality.'[13]

Nevertheless, with the science of algorithms the twentieth century experienced a new kind of abstraction, the abstraction of *virtual realization*, the study of abstract theories that make possible the processes of computation that take place in the physical space and time of a machine. This new abstraction, noted Markov, consisted above all in distancing oneself from the real limits of our possibilities of construction, and in starting to establish its theoretical premises.[14] Actual realizability was to be considered, in the first instance, as merely virtual. In effect, in his celebrated treatise on recursive functions, Hartley Rogers declared that he asked himself 'questions of existence or non-existence of computer methods, rather than questions of efficiency or good design'.[15] In Turing's own machine, famously described in an article of 1937, one was faced with a peculiar combination of the concreteness that is typical of a mechanical contraption and the absolute mathematical abstraction that accords to an immaterial idea of calculability. A practical realization, by means of concrete procedures, would depend on the actual availability of sufficient resources of time and space. From the second half of the twentieth century a science of algorithms would be devoted to this, tasked with the role of measuring the computational cost of a procedure precisely in terms of space and time of execution.

15. The Growth of Numbers

In Greek mathematics the processes of approximation, the progression of numbers and the enlargement (or reduction) of geometrical figures posed serious questions regarding their prolongation to infinity. The difficulties were resolved by assigning to the *ápeiron* a purely potential meaning, just as in the methods of exhaustion or in interactive algorithms based on the repeated, indefinite application of an operator to the approximate value of the solution of a problem, updated step by step. In these algorithms the calculated numbers grew in correspondence with growth or decrease of the figures, and since accuracy would have required increasingly larger numbers the calculation would have become, from a certain point onwards, altogether impossible.

Moreover, in Greek mathematics large automatic calculation did not exist, and deliberations dealt with relatively small numbers. Only Archimedes, in *The Sand Reckoner*, thought about a method of conceiving of much larger numbers, of millions of millions of sums, sufficient to count enough grains of sand to fill the entire universe. But these numbers, albeit extremely large, were scarcely comparable to the infinite, and above all did not form part of real computation. Consequently, in Archimedes' treatise one cannot glean any preoccupation with their growth, and their actual existence was not questioned.

In the modern science of computation, difficulties also arise in procedures with a limited number of operations, due to the demand for a degree of actuality that the *finite*, in and of itself, is not capable of guaranteeing. In general, automatic computation with rational numbers of the form p/q, where p and q are whole numbers, has a problematic character from the most elementary phases of its operations: one cannot rule out the possibility that a few multiplications will produce enormous numbers that exceed the limits of space in the calculator's memory.

A further question arises: how closely is it actually possible to approximate an irrational number by way of a fraction? The enlargement or diminution of geometrical shapes has as its counterpart the growth of numbers. An exemplary instance is provided in the case of the approximation of the square root of a whole number with Newton's iterative method, whereby the number of sums of the numerator and denominator of the approximate fractions is doubled every time in correlation with a sequence of squares constructed with successive gnomonic corrections. The lateral and diagonal numbers also grow, albeit more slowly.

But for how long does it make sense to prolong the iteration? In principle, the numerators and the denominators of the fractions could grow indefinitely, in order to improve as far as possible the approximation of the irrational number. Nevertheless, there are objective limits to the progressive diminution of the distance between the irrational number and the approximating fraction. With a theorem demonstrated in 1844, Joseph Liouville established the theoretical limit for the approximation of an *algebraic* irrational number, that is to say, of a number that is the solution of an algebraic equation, obtained by equating to 0 a polynomial with integer coefficients: the error of approximation, that is to say, the distance between the algebraic number and an approximating fraction p/q does not diminish as much as we might desire or expect as q grows. But Liouville's theorem has a wealth of implications of a different kind – from the construction of transcendental numbers, that is to say, non-algebraic ones, to the measurement of the speed of growth of numbers during the process of approximating an algebraic number. In every case the growth of numbers is the price paid for the gradual closing in on the solution to the problem.

More precisely, a number is *algebraic* to a degree k if it is the solution to an algebraic equation obtained by equating to 0 a polynomial of degree k, but not a polynomial of a lower degree than k. For example, $\sqrt{2}$ is an algebraic number of degree 2 because it is the solution of the degree 2 equation $x^2 - 2 = 0$, and is not the solution to any linear equation (that is to say, of degree 1). Liouville's theorem states that for every algebraic number x of degree $n > 1$ the growth of q does not result in a lessening, beyond a certain limit, of the distance from x of an approximate fraction p/q. This distance remains larger than $1/q^{n+1}$ for denominators q that are sufficiently large. For example, the difference in absolute value between an irrational number defined by the square root of a positive whole number and its approximate fraction p/q remains higher, for a sufficiently large q, to $1/q^3$. The numbers that are not solutions to any algebraic equation, *transcendental* numbers such as e and π, are not subject to this limitation. There is a non-numerable infinity of transcendental numbers, because as Cantor demonstrated the algebraic numbers form a numerable set, and the real numbers, including algebraic numbers and transcendental numbers, form a set of greater power than the numerable. Thanks to his theorem, Liouville was able to demonstrate for the first time how it is possible to *construct* transcendental numbers. Liouville's transcendental numbers take the form:

$$a = 0.a_1a_2000a_3000000000000000000a_4000000000......$$

where a_i is a decimal number between 1 and 9, and the groups of 0s have a length that grows very rapidly (with a factorial speed, therefore greater than an exponential one). Now, supposing that a is an algebraic number of degree n, if one considers an approximation b of a obtained with a sufficient number (depending on n) of groups of 0s, we obtain a distance between a and b that is incompatible with Liouville's theorem. Therefore a cannot be an algebraic number. In this case, it is precisely the very rapid growth of the length of sequences of 0s that enables it to be established that a is transcendental.[1]

Liouville's theorem also makes it possible to measure the speed of growth of the denominator of the fractions with which it approximates with an iterative algorithm, an algebraic number z: the growth is all the more rapid the greater the rapidity of convergence of the method, that is to say, however more rapidly the calculated fractions get nearer to z. And the same theorem also makes it possible to find a relation between the size of the denominator and the number of operations required to calculate it.[2]

That said, the principal inconvenience of the growth of numbers is a different one. In digital calculation, numbers consist of finite sequences of binary numbers (0 and 1). But with every operation using these sequences, whose number of digits cannot exceed a fixed limit, an error of rounding is produced that can generate numerical instability and a fatal loss of the information necessary to ascertain the solution to a problem with a sufficient degree of precision. The theory of disturbance is an essential stage in digital calculus, as well as with regard to the theory of dynamic systems and of deterministic chaos. A condition for the predictability of the behaviour of a phenomenon simulated by a dynamic system which is discrete, defined by a simple iterative law, resides in its sensitivity to error. Prediction as to how a phenomenon will develop may be altogether unreliable if we admit perturbations, however small, with respect to the initial conditions from which it begins to evolve.

Computation using rational numbers of the form p/q is generally impractical. The numerator p and the denominator q usually grow too rapidly, to the extent that they exceed the limits of the machine's memory. The growth of numbers is also reflected in computational complexity: the calculation of p and q is all the more expensive the larger these numbers become. The first remedy consists in abandoning the representation of a rational number in the form p/q, and in the use of *floating point* numbers $(.a_1 a_2 \ldots a_n)B^k$, where n is a predetermined number and information on the size of the number is located on the exponent k. Only in a few particular cases will it be advantageous to resort to traditional calculation with fractions, not so much by way of simplifications obtained by dividing p and q for eventual common factors, but rather as a result of continuous approximations and special techniques of rounding.[3] This realization is not enough, however, to dispel the difficulties due to the abnormal growth of numbers or operations. Growth may consequently cause the disappearance of all purported reality from what one is trying to calculate.

As Donald Knuth writes, 'experience with fractional calculations shows that in many cases the numbers grow to be quite large . . . it is important to include tests for overflow in each of the addition, subtraction, multiplication, and division subroutines'.[4]

Mathematics does not occupy itself, usually, with calculating numbers, regardless of how small or big they might be. If it does so, it is only out of practical or applied necessity. Nonetheless, the question of *how* to do it, of *how* one might calculate the actual numbers that are indispensable for a description of the world, poses problems of notable theoretical interest. The translation of differential and integral models into algorithms and into digital computation on a large scale, which is ultimately the contemporary version of the arithmetization of analysis at the end of the nineteenth century, leads to a reformulation of the great issues faced by research into fundamentals in the first part of the twentieth century: when and in what way does a problem or a class of problems admit a solution? What does it mean to *actually* resolve a problem? What meaning does the finite and the infinite have in automatic procedures? What relation is there between the continuous and the discrete?

In the early forties, numerical computation, which consists of solving an equation in numerical rather than purely analytical terms,[5] was still in a relatively rudimentary state: it was more of an art than a science of computation. To a certain extent the situation was paradoxical, because mathematics since the Babylonian calculus of antiquity, that is to say since 1800 BC, had been based on algorithms of a purely *arithmetical* and *numerical* kind, as we find prefigured in part in the rigorously *geometric* theory of Euclid's *Elements*.

The procedures that were used to approximate irrational numbers, such as those based on *antanaíresis*, had an iterative form: the approximate values were calculated with recurrent formulas that reproduced numerically the growth of geometric shapes. Algebraically, the Euclidean algorithm was virtually linked to the construction of fractions, the so-called *continuous fractions*, capable of approximating irrational numbers, and the real numbers may be *defined* in terms of continuous fractions. But the recursive generation of the numerators and denominators of fractions often produces phenomena of uncontrolled growth that can bring a halt to computation.

The most advanced kind of computation has inherited, surprisingly, the schema of ancient mathematics, which was inspired in turn by religion; but the criteria that inform it have also imposed an analysis of the efficiency of the procedures. Ancient formulas, when of a recursive kind, were already mechanical procedures, in accordance with the most recent demands of digital calculation. But only now are we conscious of the fact that every cycle of a recursive process, however simple, can inherit the errors made in preceding ones.[6] And it is therefore fatal that even the most elementary recurring relations with which fractions are constructed to approximate irrational numbers are exposed to phenomena of numerical instability that render *unreal* the very same numerical entities that Cantor and Dedekind thought actually existed.

Also reduced to recurrent relations is the calculus of special functions, such as the so-called Bessel functions, as well as particular classes of numbers such as Bernoulli numbers. Jakob Bernoulli first introduced the latter in his *Ars Conjectandi* (1713), and Leonhard Euler subsequently studied them in greater depth. These numbers are implicated in simple discrete sums that approximate integrals and consequently in questions regarding the relation between the continuous and discrete, between arithmetical calculations and continuous phenomena, between numbers and geometric magnitudes, that have marked mathematical research ever since Zeno's paradoxes. Their properties, as well as their singular relation with the number π, justify a separate study of considerable theoretical interest. Bernoulli numbers are rational numbers, fractions of which the denominators are explicitly known but not the numerators, which increase rapidly. In order to calculate them, in recent periods complex algorithms have been devised: two of the five prime factors of the numerator of B_{200}, the two-hundredth Bernoulli number, of 90 and 115 digits respectively, have been evaluated thanks to the most powerful digital calculators. Euler must already have been astonished by the speed at which Bernoulli numbers grow. These form, he remarked, 'a sequence that diverges rapidly and grows with greater force than any other geometrical sequence of increasing terms'.[7]

16. The Growth of Matrices

Every problem in applied and computational mathematics can be traced back in the end to the aim of resolving a system of *linear* equations. This is a thesis, or rather an observation, that is by now widely accepted. If the problems are not linear, they become linearized, and in this way we derive, in an approximate way, the first useful information on the solution of a differential or integral equation, of a problem of the minimum or of polynomial approximation. The approximation is not a flaw, it is the norm, and everything depends on quantifying it, in establishing when it is possible and convenient. The linearization implies an error in the approximation, but one not more fatal than others that are entailed in the various stages that lead from the initial equations to pure digital calculation.

From the second half of the twentieth century onwards it became clear that the systems of linear equations that were useful for applications have very large dimensions, and given that a system of linear equations is identified by a *matrix* of coefficients, that is to say, by a table of numbers arranged in rows and columns, in the decades that witnessed the earliest development of digital computation an in-depth study of matrix calculation on a large scale was urgently required. Now this was the principal difficulty: to reconnect the initial model, that is to say, the differential or integral equation, with the purely mathematical calculations necessary to solve a system of linear equations. The term 'linear' should not deceive us. The distinction between linear and non-linear has often marked, in computational and applied mathematics, a critical border of resolvability, but there are *linear* problems that are absolutely resistant to any attempt at resolution. This is typically the case when the matrix of coefficients of a system of linear equations is strongly ill conditioned, that is to say, when there is a high awareness of the solution of the system in relation to errors in the data.

A *matrix* A of n rows and m columns is a table of numbers arranged in a rectangle, where the element of position (i, j), consisting of the row i and column j, is denoted by the symbol a_{ij}. If $n = m$, the matrix has a square form. Matrices are self-sufficient mathematical entities, just like numbers, polynomials and functions, and the algebra of matrices is based on sum, product and inversion operations, analogous to those between numbers but usually devoid of certain properties, such as commutativity of the product, that apply to numbers. The elements of a matrix may be the coefficients of a system of linear equations, while the known terms of the equations are arranged in a column of numbers that defines a *vector*.

In current symbolization a vector is usually represented as an arrow, with its own direction and length, but it is also possible to conceive of a vector as like a column of numbers. On the Cartesian plane a vector v with one may be represented with a column of only two numbers x and y, corresponding to its projections on the axes.

In a space with n dimensions a vector v is consequently defined as simply a column of n numbers. The number of dimensions does not necessarily presuppose an idea of space or space-time. To introduce more dimensions may simply be a *way of thinking about the problem in mathematical terms*. In economics it may be necessary to configure an expenditure divided into industrial sectors. Given, let's say, ten sectors (automobile,

Figure 8

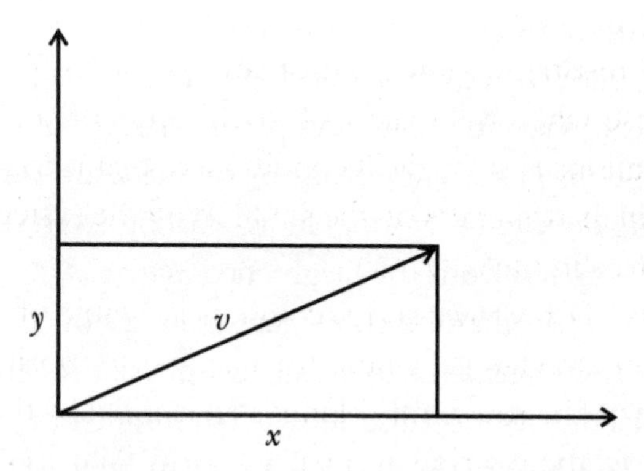

textile, naval, construction, etc.), the expenditure will be represented by a vector of ten components, each one of which is the number that measures the cost in a particular sector.

A square matrix A acts on a vector v like an operator or a transformation that associates v with another vector w. We then write $w = Av$, with w being equal to the product of A and v.[1] The case then becomes important in which, under the effect of a matrix operator A, the vector v does not change direction, that is to say, it is transformed into a vector w that has the same components of v multiplied by the same real or complex number λ. We thus write this as $Av = \lambda v$, a formula that expresses a kind of *invariability*, because the direction of the vector remains unchanged under the effect of the operator A. In which case λ is an *eigenvalue* of the matrix A, while v is the corresponding *eigenvector*. It is the eigenvalues of a matrix, in many cases, that decide the degree of efficiency of a method for resolving a system of linear equations $Ax = b$ (where b is the vector of the known terms), as well as awareness of the solution x with respect to the errors for the data A and b.

The algebra of vectors and matrices was initially developed in the context of geometry and algebra, but over time the implications of matrix calculation became ever more evident in different sectors of applied mathematics, in the study of equilibria, mechanics, theoretical physics, electronic engineering and computing, for example in interactions with the numerical-computational aspects of online search engines.

The true novelty of computation, starting from the forties and fifties, derived from the heightened *dimension* of problems and from the fact that the coefficient matrices of a system of linear equations that aims to simulate a physical phenomenon with sufficient precision has tens of thousands of rows and columns. The resolution of such a system hence requires an automatic procedure that takes into account the techniques of the *representation* of numbers, the *time* taken to execute and the *space* of memory that is required. This demands a radical change in the arithmetical laws with which calculations are made.

The algebra of matrices became in turn a central chapter in digital calculation. To linear algebra and matrix calculation almost all the computational problems of mathematics can be traced.[2] Gilbert Strang's words on this topic are extremely telling: 'For engineers and social and physical scientists, *linear algebra* now fills a place that is often more important than calculus. My generation of students, and certainly my teachers, did not see this change coming. It is partly the move from analog to digital, in which functions are replaced by vectors. Linear algebra combines the insight of *n*-dimensional space with the applications of matrices.'[3]

In the first phases of the development of calculus on a large scale, during the forties, a standard manual that was frequently consulted and went through several editions was Whittaker and Robinson's *The Calculus of Observations*. But the text was largely concerned with problems of interpolation, on the resolution of non-linear equations, on numerical squaring and on Fourier analysis. The resolution of a system of linear equations did not figure among the problems analysed per se. It was only a functional appendix to a theory of approximation based on the technique of least squares, and the only technique that was recommended, and described in a few pages, was the one devised by Cramer. The utterly ineffectual nature of this method was soon enough made abundantly clear.

Whittaker and Robinson's manual was also consulted by James Wilkinson, one of the pioneers of the science of calculus in the twentieth century, who worked alongside Alan Turing at the National Physical Laboratory between 1946 and 1948. During this period, in the Armament Research Department, Wilkinson was given the task of solving twelve linear equations with twelve unknowns – and his initial attempt, as he describes it himself, made use of the Cramer method. This was perhaps the first time in which a recursive method, with a *finite* number of operations, revealed itself to be absolutely unattainable. The cause of this was due not to the problem but to the method, which requires a factorial number of arithmetical operations and hence a calculation time, for only fifty equations – with about 10^{-6} seconds to execute a multiplication – that is greater than the time between the birth of the universe from a hypothetical Big Bang to the present day. It was evident that a solution that would take such astronomical albeit finite time to calculate was no solution at all, and that it was in a certain sense completely *unreal.* The reason for this was the increase in the number *n* of equations, the abnormal rather than exponential growth of the quantity of arithmetical operations.[4]

In 1894 David Hilbert published an article on approximation, using the criterion of least squares, of a continuous function by a polynomial.[5] The problem confronted by

Hilbert corresponded to a fundamental computational strategy: to bring back the evaluation of any (continuous) function to a simplified model of calculus, such as that of a polynomial, consisting exclusively of additions and multiplications. Mathematics has always sought such simplifications, ever since Newton's algebra, which aimed to represent functions $f(x)$ of notable analytical complexity with simple combinations of powers of x. In the nineteenth century various mathematicians had devoted themselves to this topic before Hilbert, and in 1885 Weierstrass had demonstrated that every continuous function defined in a closed interval allows itself to be uniformly approximated by a polynomial. Therefore, except for an error of approximation that is in any case inevitable, the calculus of the continuous function was reduced, at least in principle, to a finite number, dependent on the grade of the approximating polynomial, of sums and multiplications only. This could also be accommodated in the more extensive project of arithmetizing analysis.

Now Hilbert's problem required the resolution of a system of linear equations in which the matrix H of the coefficients, square with n rows and n columns:

$$\begin{bmatrix} 1 & \frac{1}{2} & \frac{1}{3} & \cdots & \frac{1}{n} \\ \frac{1}{2} & \frac{1}{3} & \frac{1}{4} & \cdots & \frac{1}{(n+1)} \\ \cdots & \cdots & \cdots & \cdots & \cdots \\ \frac{1}{n} & \frac{1}{(n+1)} & \cdots & \cdots & \frac{1}{(2n-1)} \end{bmatrix}$$

which is known today as the model of a numerically intractable matrix. The matrix H is now called the *Hilbert matrix*.

One of the fathers of computer science in the last century, George Forsythe, had to deal with the Hilbert matrix several decades later. In an important article of 1970 he observed that, for a dimension n barely greater than 8 or 9, any automatic procedure does not have the capacity to solve a system of n linear equations with n unknowns defined by H, and that for small values of n, such as $n = 6$, there was a large discrepancy between the solutions of apparently identical problems calculated on machines that were just slightly different. We would now argue that the reason for these difficulties resides in the fact that the matrix is very *ill conditioned*. Now, a problem is described as ill conditioned if small variations in the input data can produce great variations in the solution, regardless of the procedure used in the calculation.

The procedure of approximation studied by Hilbert consists of making minimal an expression of the error due to the substitution of a continuous function by a polynomial. To minimize the error we must then solve a system of linear equations $Hx = b$, the matrix H of which is intractable even for small dimensions.[6] Already for $n = 6$ the elements of the inverse H^{-1} of H

$$\begin{bmatrix} 36 & -630 & 3360 & -7560 & 7560 & -2772 \\ -630 & 14700 & -88200 & 211680 & -220500 & 83160 \\ 3360 & -88200 & 564480 & -1411200 & 1512000 & -582120 \\ -7560 & 211680 & -1411200 & 3628800 & -3969000 & 1552320 \\ 7560 & -220500 & 1512000 & -3969000 & \mathbf{4410000} & -1746360 \\ -2772 & 83160 & -582120 & 1552320 & -1746360 & 698544 \end{bmatrix}$$

lead us to foresee their immeasurable growth, even for values of n that are just slightly greater.[7] Now, the number 4410000 in the position (5, 5) of the inverse of H has the capacity to amplify a perturbation in a catastrophic way, even if it is very small, on the fifth element of vector b of the known terms.[8]

For $n = 6$, the number 4410000 is the element of maximum modulus of the matrix H^{-1}; but if we allow n to grow, the errors regarding the elements of x become ever greater. For $n = 10$, the maximum element, in modulus, of the inverse is 3.48×10^{12} and therefore, if there is an error present in some element of b, there would be a disproportionate degree of error in some element of the solution x. The matrix condition number climbs to around 10^{13}.

All of this has to do with the structure of the mathematical problem and does not depend on the choice of the algorithm that has been delegated to solve it. But there is also a degree of sensitivity of the system of calculus with regard to the errors made in single operations of the algorithm. Realizing that Cramer's method for solving a system of twelve equations was completely inefficient, Wilkinson chose to use Gauss's algorithm, which consists of reducing to 0 all the elements of the matrix of the coefficients under the principal diagonal, thus reducing it to a matrix T of triangular shape (with the elements beneath the diagonal equal to 0). Gauss's method is incomparably more efficient than Cramer's, but it is not always able to dispel an abnormal growth in the numbers that are progressively calculated. On this growth the degree of error in the calculus depends, and a formula that establishes a limit superior to the error as a function of the magnitude of the elements of the triangular matrix expresses this dependence with deterministic precision.

Usually the error made with the Gauss method is limited at the upper end with an expression E that consists of a measure of the *magnitude* of the matrix of coefficients A multiplied by the square of n and by a *factor of growth*. This factor is defined precisely by the relation (in absolute value) between the maximum of the elements calculated for the triangular matrix T and the maximum of the elements of A. The result does not have a probabilistic significance but a deterministic one, in the sense that for every possible value of the dimension n of the matrix, the error can never be greater than E.[9] We may therefore conclude that when the calculated values of T become too big, the entire computational picture loses all sense. The factor of amplification of the error may be expressed in a variety of ways, but always ends up by depending on the growth of the numbers that have been actually calculated.

In 1946, Hotelling fixed a limit to error of an exponential nature, proportional to 4^n, that left little hope for the possibility of resolving the problems of mathematical physics numerically with automatic procedures. Nevertheless, the initial pessimistic predictions were to be rectified by more accurate analyses. In 1947 John von Neumann and Herman Goldstine published a historic article in which they analysed the behaviour of the error in the inversion of a positive definite symmetrical matrix,[10] with the Gauss method, by means of an arithmetic with a predetermined index of precision. Von Neumann and Goldstine demonstrated that the error, however subject to fluctuations that could not be controlled directly, does not exceed a fixed amount that depends on the dimensions of the matrix, on the number of digits with which the calculation is executed, and on the relation between the maximum and minimum eigenvalues of A.[11] It amounted to a result that could be called somewhat miraculous: for large dimensions of the matrix, and for millions or billions of approximate operations, it could seem necessary to resort to a *statistical* analysis of the errors of rounding (which was in any case carried out); but the von Neumann and Goldstine result had a *deterministic* character, even though within the confines of an interval of indeterminacy: the error, in itself unknowable, would not *in any case* have gone over a certain threshold that was, at least in principle, perfectly capable of being evaluated and dependent on the growth of calculated numbers.

In 1971, in an important article surveying the analysis of error elaborated in the preceding twenty to thirty years, Wilkinson was prompted to comment that it was inaccurate to assert that the error accumulated in certain stages of the calculation is *amplified* in subsequent stages, and that in fact, if the number of operations is sufficiently high, the errors are compensated for statistically – so much so as to diminish, paradoxically, their effect.[12] The most dangerous enemy is not the propagation of error through the operations, but the error *inherent in the problem* of the inversion of the matrix, that is to say, the degree of intrinsic sensitivity of the inverse of A with respect to the small variations of the elements of A, regardless of the choice of methods adopted. This sensitivity is measured by a *condition index*.

The concept of the condition of a matrix was specified for the first time by Alan Turing in his 1948 article on the numerical solution of a system of linear equations and on the inversion of a *general* matrix, that is to say, not necessarily symmetrical and positive definite, with elements equal to real numbers.[13] Turing introduced a *norm*, a real number associated with the matrix, an index that was meant to express the *magnitude* – not the dimension, given by the number n of rows and columns, but the overall magnitude of its elements. Turing's chosen norm, today known as the *Frobenius norm*, was defined by the square root of the sum of the squares of the elements of the matrix: a simple *number*, therefore, to which the task of measuring the magnitude of an entire *matrix* was delegated. Now, Turing remarked, it is possible to express the error arising in the solution of a system of linear equations as the error in the elements of the matrix A of the coefficients multiplied by the factor μ equal to the product of the norm of A and the norm of the inverse of A (divided by n). The amplification of the error that is present in the elements of A was therefore seen to depend on the magnitude of this factor, which as a result took the meaning of the *condition number*. If μ is high, we will not know if the calculated numbers will give plausible information on the exact solution of the linear system, which is equivalent to saying that we will not have sufficient information on the solution of the equations, of which the linear system is an arithmetical approximation that defines the mathematical model of the physical phenomenon the evolution of which we are interested in predicting. Predicting the progress of this evolution is instead possible if μ is relatively small.

These considerations do not rely on the algorithm selected to solve the system of linear equations: the condition number depends only on the intrinsic properties of A and its inverse. Nevertheless, the index μ also intervenes in the stability of algorithms, namely, in relation to the actual procedures by means of which the system of equations is solved numerically, that is to say, the way in which the error propagates itself through the operations. The measure of the sensitivity of the result of the computation with regard to errors in the operations (necessarily approximated) may be summarized in an expression that depends essentially on μ.[14] This circumstance was already implicit in the 1947 article by von Neumann and Goldstine on the inversion of symmetrical and positively defined matrices. In this article the error did not exceed a value dependent on a factor μ capable of being expressed as a relation between the maximum and minimum eigenvalues of A; but this relation coincides with the relation between $N(A)$ and $N(A^{-1})$, where N denotes a norm, the Euclidean norm that is equivalent but not identical to the Frobenius norm used by Turing.[15] In the specific case of positive definite symmetrical matrices considered by von Neumann and Goldstine, the Euclidean norm coincides with the maximum eigenvalue of the matrix. In such a case, then, it is the same maximum eigenvalue that measures numerically the *magnitude* of the matrix.

To conclude, the overall error does not exceed a value that depends on the way in which numbers grow during the computational process. If these numbers become very big, every aspect of predictability of the mathematical model is impaired, together with all precision. It is a question not merely of the loss of information due to the rounding of very large numbers, or of the risk of *overflow*. The crucial fact is that the growth of calculated numbers, increasing the condition index, impedes prediction of the error's behaviour. This index acts as an essential component of the expression that defines the upper limit of error, a limit that cannot be exceeded. If the limit is a high number, the error *might* be big,[16] and it is precisely this situation of uncertainty that puts into doubt and renders inscrutable the meaning of numbers printed at the end of the process of computation.

It is not just the growth of numbers that undermines a

system of computation; their decrease does as well, and the consequent operations for numbers close to 0 can produce intractable results, ending with the arrest of the computation process. In all critical cases the solution becomes evanescent, in a sense completely unreal. There are theorems that posit, in theory, their existence, but nobody is in a position to be able to calculate the figures.

What is certain, contrary to what Cantor and Frege maintained, is that numbers do not all share the same ontological status. Numbers that exist but cannot be calculated do not have the same reality as numbers calculated by a machine. The former, unlike the latter, are not situated in the space and time of an actual automatic elaboration – not actually or virtually in fact. From a certain point of view, a number exists, is real, only if there is an actual procedure that calculates it. But this procedure must also be *efficient*: otherwise, as in the case of Cramer's method or Hilbert's matrix, one would not know how to distinguish, on the level of *actual* realizability, between the calculable and the non-calculable. The calculable resembles that which is not calculable.

Boethius had explained that, however different from each other, numbers are not made up of anything other than themselves.[17] The concept of the closure of a numerical field F would go on to guarantee that, by combining numbers with the operations defined in F, one does not find elements alien to it but always and only elements of F. But what can one say about numbers calculated by a machine? In this case, the operations on the numbers always generate other numbers; but these, during the course of computation, might lose all meaning. The algebraic closure of the field is not sufficient to guarantee the significance of numbers generated in a computational process.

The concept of a whole number is not in itself capable of explaining in all its generality the concept of algorithm as an actual process that unfolds in time and space. Dedekind had sought an affirmative response, but had only managed to demonstrate that arithmetical recursion, limited to the operations of addition, multiplication and division, leads back to set properties that allow the concept of natural number to be founded.

What do we mean, exactly, by 'effective'? The term is often used in the literature but is hardly ever defined in an explicit way. Andrej A. Markov Jr attempted to explain that the effectiveness of an algorithm is one of its essential prerogatives, consisting of the 'tendency of the algorithm to obtain a certain result',[18] calculated on the basis of a set of initial data. As a result, it became standard to call actual every process of calculus defined in a recursive way (a criterion equivalent to λ-formalism or Turing's machine), on the grounds that the recursion, already as early as Dedekind's theory in the nineteenth century, is the technique for constructively defining the fundamental operations of arithmetic. But in parallel with the effort to define it theoretically, the algorithm also became the protagonist of a computational mathematics oriented towards resolving problems of applied science. In this context, at least from the second half of the twentieth century onwards, the accent began to shift from effectiveness to a more radical and operative demand for *efficiency*.

The fundamental blueprint has always consisted of the arithmetization of mathematics, of the reduction of analysis to the concept of whole numbers and operations in the passage to the limit. But in order to realize this reduction, efficient algorithms are required, both to establish complex computational strategies and in order to execute the elementary arithmetical calculations *hidden* within these strategies, such as the additions and multiplications between numbers, the linear interpolations, the products of polynomials or matrices. It is worth repeating: the calculator executes millions of operations and makes use of numerous subprograms that are not readily accessible. Faced with the phenomenon of this immanent, concealed computation, one must be able to rely on an arithmetic of a machine that simulates that arithmetic exactly, in the most efficient way possible, in terms that relate both to stability and to the saving of time and space.

Mathematics always and above all pursues a principle of economy: the search for the shortest route to obtaining a predetermined objective. This principle is not necessarily dictated by the criterion of what is useful and adheres above all to a strategy of theoretical simplification that tends to compress the maximum information into a process with minimum labour spent on calculation. If we want to predict the behaviour of a phenomenon, we must prevent the information contained within mathematical models from being translated into arithmetical detritus without meaning: the lists of numbers that are obtained at the end of the procedure must reflect, at a lower level of information, in the time and space of the calculator, that which the model intended to represent. The efficacy of mathematics in representing the physical world is transmitted along the whole complex and bumpy intermediate road between the theoretical model and the numbers that are actually calculated. In the course of this passage we can also lose direct contact with the nature of the physical phenomenon the evolution of which the initial model sought to simulate, because the efficiency of the calculation ends up ultimately being founded on pure mathematics, on theorems and mathematical structures that have little to do with physics, the efficacy of which nevertheless produces results that are no less miraculous than the initial correspondence of the theoretical model with the nature of the phenomenon. This obtains especially in the case of matrices: if, as usually happens, a differential model comes into proximity with a system of linear equations, the matrix of the coefficients of the system has a structure that reflects that of the model, but a solution may involve matrices of a completely different structure.

In 1946 Herman Goldstine and John von Neumann wrote the following on the subject of the large-scale digital calculation that was beginning to be delineated at the time, with a particular bearing on the mathematical solution of problems of fluid dynamics by means of differential models:

> It is important to avoid a misunderstanding at this point: one may be tempted to qualify these problems as problems in physics, rather than in applied mathematics, or even pure mathematics. It should be emphasized that such an interpretation is wholly erroneous. It is perfectly true that all these phenomena are important to the physicist and are usually mainly appreciated by him. Yet, this should not detract from their importance to the mathematician. Indeed, we believe that one should ascribe to them the greatest significance from the purely mathematical point of view.[19]

Nevertheless, calculus must be understood down to its last details. The great theoretical syntheses provided by philosophy have also sought, in compliance with a principle of reality, a precise correlation with the most elementary operations of computing. In this sense, mathematics is held to be reductionist, but the reductionism must not encourage us to forget that all the intermediate stages between the model and the numbers actually calculated are possible only thanks to the intervention of theoretical concepts and relatively abstract mathematical structures. In an expanded idea of reality the most abstract structure counts just as much as the list of numbers – which by itself is unintelligible – that a calculator prints materially at the end of a process of calculation: they are mirror images of each other, with each being decrypted by means of the other. Therefore, mathematics is not just an abstraction. But precisely because of its abstract nature and its theoretical content, it is a foundation of the reality of this world and of the way in which we venture to intervene in order to modify it.

The proximity of mathematics to that which we are prepared to qualify as real and actual evokes the nature of that power which the Greeks assigned to the simplest operations, such as constructing a square above a line, and which Plato has countersigned with the term *dýnamis*. Already back then potency (*dýnamis*) was capable of producing, besides the growth of temples and altars, the most sophisticated mechanical contraptions, catapults and other war machines. It was therefore inevitable that the Platonic *dýnamis* could transmute itself into actual scientific and technical power, with all the risks that we can now imagine and counteract. A clear sign of the affinity between reality and power, which may be extended to the entire scope of nature, is to be found in Spinoza's *Ethics*, in a period already profoundly marked by the Promethean leap of science: 'nature is always the same, and its power to act is everywhere.' And in conclusion: 'The power or force of anything . . . is nothing but the given, that is to say actual, essence of the thing itself.' (*Ethics*, III, Preface and Proposition 7, Demonstration.)

17. The Crisis of Fundamentals and the Growth of Complexity: Reality and Efficiency

It is not just the numbers that grow, the complexity of calculus also increases. And this insidious feature may also threaten the best-known and most-tested procedures. It is not only Cramer's algorithm that is affected. If we employ the classic Gauss method to resolve a system of linear equations or to calculate the *determinant* of a matrix,[1] a flaw in efficiency may arise from the increase of the size of the calculations: in the case of the determinant, the number of *arithmetical* operations is a polynomial function of the dimension n, but it is not always easy to verify that the number of *binary* operations also remains polynomial as regards the elementary operations on the single sums of the representations in base 2 of the numbers that intervene in the calculation. This number may become exponential.[2]

The study of computational complexity, that is to say, of the criteria for measuring how difficult it is to calculate a function, has been one of the most important streams of computer science since the fifties. With the increase in dimension, the difficulty presented by the resolving of relatively elementary problems can grow in an exponential way, due to an uncontrollable combinatorial explosion. This is the case with the so-called NP-complete problems (solvable in Polynomial time with a Non-deterministic machine): it is relatively easy to *verify* if a given solution proposed for any one of these problems is actually the case, but the *actual search* for a solution requires, according to current knowledge, a number of operations that is at least exponential.

The theory of complexity also contributed to the galvanizing of the crisis of fundamentals that had threatened the mathematics of the first decades of the twentieth century, opening new perspectives for research in areas that had been completely ignored until the forties. This revaluation was able to develop thanks to the affinities between the problems tackled, as well as between the techniques that were necessary to clarify them: the central question that initiated research into fundamentals, from the celebrated list of twenty-three problems posed by Hilbert in Paris in 1900, pertained to *whether any mathematical problem can be solved*, whether in a positive sense (with the actual calculation of the solution) or in a negative one (by demonstrating that no algorithm exists which could calculate the solution). The central question relating to the complexity of calculus is *whether a mathematical problem can be resolved in polynomial time*, that is to say, with a number of operations equal to a polynomial function of the size of the problem, and hence compatible, in the final analysis, with the time and space of the calculator's memory.

In a letter from Princeton dated 1956, Gödel put to von Neumann a mathematical question that was destined to become one of the fundamental (and still unresolved) problems of the science of calculus: we may easily construct, argued Gödel, a Turing machine that, for every formula F of the calculation of the predicates of the first order and for every fixed natural number n, allows us to decide whether there is a demonstration of F of length n (if the length n is the number of symbols, one need only specify the list of all the demonstrations of length n that can be executed by the controlling machine, then check if among these there is the demonstration of F). Now, if p is the number of steps (depending on F and n) that the machine requires for this objective, and if $P(n)$ is the maximum between the numbers p when varying F, *how quickly does* $P(n)$ grow, when n increases, for an optimal machine?

Gödel thus posed a question that did not concern pure algorithmic solvability alone but also computational complexity, as well as the measurement of the difficulty of the calculation. A comment on the letter made by Juris Hartmanis helps to better clarify its strategic importance within the field of research on the foundations of mathematics. From Gödel's results, Hartmanis explains, we know that within mathematics there is such an abundance of formulas and problems that it (mathematics) is incapable of axiomatic resolution in a coherent and complete sense. From Turing and Church's results we also learn that for formal systems that are sufficiently complex the general question of which assertions are or are not demonstrable is not something that may be decided through an algorithmic process. In his letter Gödel then raised the next question, which still has to do with the foundations of mathematics, but now in terms of the *growth in complexity of calculus*: with the increase of n, how difficult is it to decide with a procedure whether, in a formal system, an assertion admits of a demonstration of length n?[3]

Gödel, who had destroyed the dream of an axiomatized mathematics, was now hazarding, for ironic effect, the more optimistic hypothesis that the complexity of the problem posed by von Neumann could be linear or, at most, quadratic. We now know that this belongs to the class of NP-complete problems. The study of this class has required techniques altogether akin to those used for questions of pure theoretical solvability which had made it possible to reply (negatively) also to some of the queries raised by Hilbert in his celebrated 1900 lecture in Paris.

The more theoretical questions on computational complexity would be intertwined, in several respects, with the measure of efficiency of the numerical algorithms entrusted with the task of resolving various kinds of applied problems. The modalities of growth of the dimensions, of the number of operations, of the numbers actually calculated or of algorithmic error or what is due to ill conditioning are in one case or another a criterion for establishing, if not reality, then the actual existence of solutions. But the new questions, even though fundamental and partly unresolved, did not provoke a crisis comparable to that which affected mathematics in the twenties. The urgency seems now to have shifted to problems relating to theoretical computer science and the application of computational mathematics; to a comprehensive science of calculus that is in itself sufficiently empirical and flexible to tolerate operational imperfections or uncertainties.

The efficiency of an algorithm depends above all on two elements: its complexity and its stability, which is to say, its overall sensitivity with regard to errors in operations. To speak of the solution of a mathematical problem is justified and inevitable, but in the majority of cases this solution is not known at all and the calculation of its analytical formula, which is supposed to exist, would entail errors of approximation greater than those that follow from the substitution of the initial model by a simpler one. By substituting the differential or integral model with an arithmetical model the solution is then calculated with only the fundamental operations on numbers: additions, subtractions, multiplications and divisions. Nevertheless, the knowledge that may be afforded is always approximate, and the calculated results may be completely unviable. So, in conclusion, the degree of certainty intrinsic to the elementary arithmetic of numbers, on the model of which Hilbert constructed his (metamathematical) techniques designed to justify the mathematical use of the infinite, was not entirely defensible. Even one single addition, executed with a calculator, can generate numbers devoid of meaning.[4]

A typology of error pertains to computational processes of an iterative type: with repeated applications of an operator one generates a succession of numbers the distance of which from the solution (which can be interpreted as error) tends towards 0. From similar processes of calculation, the analytical concepts of the limit and the convergence of a sequence were derived, and certain demonstrations of convergence are still based on the existence and properties of this same processes of calculation rather than on logical arguments. But are we actually capable, with an iterative procedure, of getting ever nearer to a solution in a way that closes the gap (of error) to an arbitrarily small value? The mathematicians took a long time getting round to asking themselves this question, and today, in the same words as used by Cauchy, we can still hear how it is possible to calculate for the root of an algebraic equation 'approximate numerical values that are *arbitrarily* close'.[5] Nevertheless, this indefinite approach towards the solution is valid only in theory, and the error of approximation *cannot* become arbitrarily small. The errors of rounding create, fatally, around the root of the equation, an *interval of uncertainty* that divests the calculated numerical values of all meaning beyond a certain limit of approximation: a good algorithm may furnish, after a certain number of steps, a value within that interval, but it makes no sense to look to improve that value by calculating another, successive one within the same interval.

As already mentioned, in order to explain what the continuum consists of, Poincaré asked us to consider two intersecting lines. We can imagine them drawn on the Cartesian plane. Our imagination tends to visualize them as two thin ribbons which, on intersecting, are obliged to have a part in common. The mathematician takes a further step: without rejecting the suggestion of the imagination, he tries to conceive of a line without width and a point without extension. It then occurs to him to think of the line as a *limit* to which a ribbon tends as it is progressively reduced in width. The limit to which the area of intersection between the two ribbons tends as their width moves towards 0 is then a point. This is why, Poincaré concluded, we are inclined to maintain, in compliance with an intuitive truth, that two intersecting lines *must* have a point in common.

What happens if the abscissa α of the point of intersection P of the curves does not correspond to a rational number? With the strategy suggested by the theoretical definitions of the field of real numbers we try to calculate a sequence of intervals of decreasing length that include α, each contained within the preceding one. In theory, we are entitled to assume that we can define an interval of length containing α that is *arbitrarily* small. But if we use the best-known numerical algorithms designed to identify these intervals, step by step, we realize that it is not possible to reduce their length beyond a certain limit. The reason for this impossibility resides in the need to arrest or round off the sequences of numbers that have actually been calculated.

If the two curves are defined on the Cartesian plane by two functions $f(x)$ and $g(x)$, the equation that must be resolved numerically is $h(x) = 0$, where the function h is defined by the difference, for every x, between $f(x)$ and $g(x)$, that is to say, $h(x) = f(x) - g(x)$. The theoretical existence of the number α which cancels the function h is guaranteed by the celebrated theorem of Bernhard Bolzano: if in a closed interval $[a, b]$ of extremes a and b, that is to say, a set of real numbers x for which $a \leq x \leq b$, a continuous function that has a positive sign for some values of x and a negative one for others, then there is a number α included between a and b for which the function is annulled (in Fig. 9 the function h has positive sign in a and a negative in b). The strategies for calculating α frequently resort to this theorem: one calculates extreme intervals ever closer, so that the function h changes sign in each of them. The indefinite sequence of these intervals defines the number α, which exists thus, in principle, in the body of real numbers. But what can one say of the actual calculation of numbers with which one approximates α? This calculation implies, typically, the evaluation of the function h at different points, in order to determine each time if it is positive or negative. But due to errors of rounding we do not calculate $h(x)$, but rather a function $t(x) = h(x) + e(x)$, where $e(x)$ is a disturbance that does not exceed a certain positive number ε. Now if $[c, d]$ is the maximum circular interval around α, inside which the function h is in absolute value less than or equal to ε, it is not difficult to demonstrate that if $t(x)$ is positive (or negative) in that interval, it is *not* necessarily so for $h(x)$ as well.[6]

From the moment in which x is calculated to fall in the interval $[c, d]$, our calculations to verify whether that function h assumes different signs do not lead to any conclusion. This

Figure 9

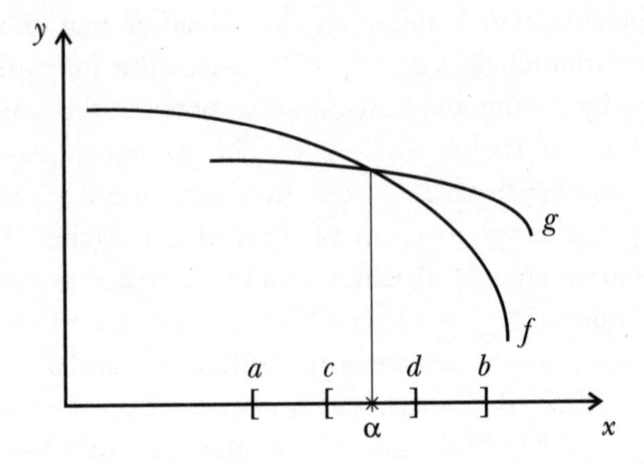

implies that it is not possible to calculate an interval smaller than $[c, d]$ that contains within it the solution α of the equation $h(x) = 0$, that is to say, the abscissa of the intersection of the two curves. The location of x within an interval that cannot ultimately be reduced is ignored.

Similar reasoning is naturally valid for more complex problems than the one described, such as the solution of a system of non-linear equations or a problem involving the minimum. Hence the cloud of indetermination that regularly surrounds the solution of a problem renders problematic its very existence. All that we know about that problem lies in the algorithm designated to approximate it, and it is understood that this algorithm must be capable of giving information about the solution adapted to the objectives for which it is being calculated. In a sense, the solution *is* the algorithm, but this also requires the *efficiency* of the algorithm, measured in terms of stability (sensitivity to errors) and computational complexity. Hence algorithmic constructability and efficiency are both requisite when establishing an ontology of numbers. Efficiency is a criterion of computational mathematics and of large-scale automatic calculation that completes the pure idea of effectiveness introduced in the context of theoretical computability.

This conclusion seems to shift the reality of numbers from a complex of theoretical definitions to an idea of concrete algorithmic efficiency. That said, for an algorithm to be efficient one needs to insert into the most concrete calculation mathematical entities which may be referred back to theorems and to mathematical structures that are relatively abstract and often distant from the model with which one aims to simulate a phenomenon. For example, the matrices that intervene in a procedure – with the sole aim of simplifying calculations – usually have a special structure, which may be described in abstract terms, in certain cases, by means of the properties of an algebraic group. Only because it entails matrices with *structure*, and not generic matrices, is it possible to bring to a conclusion calculations that would otherwise be impossible. Therefore, the materiality of efficiency presupposes the existence of mathematical entities. Goldstine and von Neumann, as has been pointed out, had already warned that digital calculus is in large part based upon pure mathematics.

The thesis, already advanced in the past, that numbers have the same ontological status as long as they are defined in a coherent way, is no longer plausible – but it is equally the case that their concrete existence depends on a theoretical knowledge that establishes the conditions and limits of their calculability.

18. *Verum et Factum*

Verum et factum convertuntur ('the true and the created are interchangeable'): this is the formula that Giambattista Vico found significant evidence of in Latin literature, in Terence and in Plautus, and which today can still serve as a premise for constructivist epistemologies. The reality of something depends on the 'doing', on actually bringing it to term with an action. The solution of a mathematical problem depends, then, on the possibility of calculating it in an efficient manner in the physical space and time of an automatic execution, which is the only strategy possible, given the size of the problems. There seems to be nothing more certain than a procedure which, through a finite number of steps, carries out the necessary calculations in relation to assigned data. But is the reality of mathematical entities really encapsulated, in a comprehensive way, in this conclusion?

It is necessary to remember that all analysis and its function as a model for the description of the physical world are based on the metric properties of the field of real numbers defined by Cantor and Dedekind. This field of numbers, which expresses arithmetically the concept of continuity, has among its peculiar qualities that of *completeness*. This means that any *fundamental* sequence of real numbers converges with a real number, that is to say, it admits a limit that is not outside but belongs to the field of real numbers. The field of rational numbers is not complete, because there are series of rational numbers that allow a limit that is not rational – and this is the case for lateral and diagonal numbers, the relations of which converge with $\sqrt{2}$, which is not rational. Now, one may frequently demonstrate the *existence* of a solution to a mathematical problem, such as an algebraic equation, an integral or a differential equation, by having recourse to the property of completeness: one calculates a sequence of values that approximate the solution and demonstrates that this sequence is *fundamental.* For the property of completeness, the limit of this sequence *exists* in the field of real numbers.

Up to this point we merely admit the existence of the solution, but in order to do so we use a sequence of values that approximate it and that are actually calculable. The demonstration of the existence of the solution resorts to an algorithm that calculates the successive approximations. Hence the theoretical property of completeness that characterizes, along with others, the arithmetical continuum, is applied to physical problems described by differential models which one seeks the solution to with algorithms that must satisfy a precise property of convergence. The execution of the calculations must be automatic, due to the elevated dimension of the problems, up to the final evaluation of interminable lists of numbers. These lists, like the procedures designed to calculate them, may not by themselves be intelligible; nevertheless, they enclose, in terms of pure sequences of digits, the information communicated by the model and hence by the same physical phenomenon which the model is seeking to simulate.

The calculation itself, ultimately, becomes a physical process, but it is difficult to separate it from the theoretical presuppositions, purely mathematical, that render it possible. The *actual* existence of the solution of a problem is certainly predicated on the *efficiency* of an algorithm, but if one attempts to grasp the reality of the mathematical entity it is difficult to separate the domains of reasoning and calculation that intersect and rely on each other, until they form a fully fledged theory – of theorems, demonstrations and algorithms, on the basis of which such reality is comprehensively supported and almost imposed. As Simone Weil writes, 'the real is that which imposes itself. The demonstration imposes itself on us more than the sensation. But it partly derives from convention. It is necessary in mathematics to capture the non-conventional.'[1] In fact, in the procedure that leads from the mathematical model to digital calculus there is little that is conventional: the nature of the physical event is necessarily transmitted in the structure of the equations and the mathematical entities designed to resolve them, until it is imprinted in the last lists of numbers calculated. Ultimately, that which has been defined as 'the unreasonable effectiveness of mathematics in the natural sciences'[2] appears to be the consequence of a complex and articulated combination of properties and circumstances that significantly attenuate the initial impression of the accidental character of the possible connections between the abstract concepts of mathematics and the physical world.

The *factum* is situated at the end of the process of modelling and materializes in the numbers actually calculated and located in the memory of the calculator; nevertheless, it is also a presupposition – historical and conceptual – of the theory of the arithmetical continuum that allows the construction and study of the model. The concept of section and Dedekind's arithmetical continuum are the natural consequence of a primordial fact that is expressed in an algorithmic construction: 'Is it not perhaps true that the concept of partition [of *section* or *cut*, in Dedekind's terms], is preceded by a purely algorithmic fact, that is to say, by the need to justify and legitimize certain algorithmic procedures, such as the approximation by excess and defect of $\sqrt{2}$, that are precisely translated into the construction of the classes made up of infinite discrete numbers?'[3]

It is possible to assert, to a certain degree, that *verum et factum convertuntur*, the true is ultimately interchangeable with the made – but we must also take into account the fact that it is difficult to make a precise definition of 'the made', or of the 'effectiveness' that should constitute the essence of a computational process. Moreover, the thesis that maintains that 'the realm of reality infinitely surpasses that of the made'[4] is certainly not inadmissible, depending on a combination of circumstances for which the made may occur in a certain way and not in another. The made, as such, remains indifferent with regard to research into its fundamentals. And it is the latter that virtually contains, as an active and productive principle, what exists in practice. As Thomas Aquinas explained, before something exists potentially there must be something that exists in practice, because potentiality does not resolve itself in practice if not by means of something that already exists in practice.[5]

19. Recursion and Invariability

The phenomenon of growth is not a marginal issue, because it is involved in the most intimate structure of the algorithm. The modality of growth has a decisive role in determining efficiency. When it is explained that an algorithm functions in a recursive way, what is meant by this is more or less the following: the problem that the algorithm is designed to solve (for instance, the product of two numbers of sixteen digits each) is divided into problems of the same kind but of lesser dimension (products of numbers with eight digits), and these are divided in turn into analogous problems that are smaller still (products of numbers with four digits). The procedure of division continues until one reaches elementary problems of minimal dimension (products of numbers with only one digit). From these latter, by way of an iterative calculus, one tracks back in the opposite direction until one solves the problem that was initially posed. This method of organizing recursive calculus, through a hierarchy of problems of the same kind but of decreasing dimension, is called *divide et impera* ('divide and conquer'), from the old Latin motto. But the criterion of division is not, in itself, always and necessarily efficient. What matters is the law that regulates the decrease and successive growth of dimensions. Cramer's method, which retraces the calculation of the determinant of a matrix of dimension n to calculation of the determinant of matrices of dimension $n-1$, is completely inefficient. A great variety of algorithms use dichotomous divisions instead, that is to say, the initial problem of dimension n is divided into problems of dimension $n/2$ (assuming that n is even). In this way, the size of the calculation follows a progression of successive powers of 2: the growth of the dimension that goes back to the most elementary initial problem doubles each time. In one respect, the algorithm recalls the theologians of Heliopolis' genealogical tree of the Ennead, in ancient Egypt, as described on the sarcophagus of Petamon (no. 1160) in the Cairo Museum:

> I am the One that transforms into Two,
> I am the Two that transforms into Four,
> I am the Four that transforms into Eight,
> After this I am One.

Division also took place in Heliopolis by means of the powers of 2, that is to say, $2^0 = 1$, $2^1 = 2$, $2^2 = 4$, $2^3 = 8$, and, after having repeated it as many times as required, the one was regained by following a path in the opposite direction, exactly as in the recursive procedure. In effect, after the reduction of an initial nucleus of small dimension, the recursive calculus consists of the progressive augmentation of the dimension which enables the solution of the problem, that is to say, the calculation of the assigned function. The procedure restarts from a small initial nucleus and becomes a growth that maintains the form unaltered. In Egypt it was not only the theological formula that had recourse, metaphorically, to the idea of progressive growth by the powers of 2. The arithmetical formulas for calculating the product of two numbers was regularly based on an analogous criterion of growth.[1]

The growth of numbers can also occur with different modalities, that is to say, not by successive powers of 2, for example, according to the celebrated Fibonacci series, in which every number different from 1 is equal to the sum of the two preceding numbers. Dividing a problem according to Fibonacci's criterion, or other, analogous successions, one may gain computational advantages, as demonstrated in the case of the parallel calculation (where numerous operations are executed simultaneously) of a rational function (the quotient of two polynomials). The working of the calculation may be described by means of a graph that unfolds in space, and in the shape of the unfolding of the graph it is possible to identify the computational complexity. The criterion of growth may differ from one case to another, but it is nevertheless decisive in defining the fastest course leading us to the final result of the calculation.

Iteration and the pursuit of invariability are the necessary ingredients, the golden rule, of computation. We can see an example of this even in the calculations which are employed by search engines on the Web. Every page or document of the Web – of the immense network of information on a planetary scale – is represented as a node i of a graph of enormous proportions, in which the number of such nodes is extremely large and to which a matrix L of equivalent proportions can be associated, with billions of rows and columns (even if we are dealing with sparse matrices, that is to say, with many 0s). The matrix L characterizes the graph, in the sense that the element l_{ij} is equal to 1 if there is a connection between the node i and the node j, and l_{ij} is equal to 0 if not. If we have to measure the importance of a node on the graph, we take into account the number of demands that converge in time on that node. This criterion may then be expressed in an iterative calculation, fixing a vector x_0 of initial hypothetical estimations of the importance of the nodes and updating it repeatedly with a new vector x by way of linear combinations of the importance values of the connected nodes – those that lead back to the node in question and those to which that node may lead. The importance of the node, that is to say, the document on the Web, depends upon the number of connections. The updating is then expressed in the iterative calculus, approximated, of the autovector x corresponding to the maximum eigenvalue of a matrix A that depends on L (A substitutes for L with the aim of making the iterative calculation convergent). Given that A has a structure that allows it to be established that its maximum eigenvalue is equal to 1, we get $Ax = x$, which is to say that x is a *fixed point* of A, a vector that remains unchanged, either in direction or length, following the application of the linear operator A. Therefore, a criterion of invariability presides over the iterative calculation of the solution of the problem of the Web, which consists in assigning major or minor importance to a page in order to elaborate a response to a generic inquiry.

The theory of matrices involved in the calculations is principally due to the genius of Oskar Perron, who elaborated it at the beginning of the last century, for reasons quite different from those for which it is famous today on account of its numerous applications to the economy, the growth of populations and models of nuclear fission. It is an irony of history that, though wholly unaware of its future applications, Perron in a letter to one of his students dated 1972 expressed all his reservations about the growth of information technology.[2] He asked himself if the latter was not synonymous with or a secondary branch of 'espionage [*Spionatik*]', with an obvious allusion to the decryption of German communications that Turing had devoted himself to during the Second World War. Another example of the unreasonable and involuntary efficacy of mathematics for the study of the natural and artificial sciences, and as an unexpected tool to counter the *hýbris* of numbers that grow immeasurably.

Notes

Introduction

1 Eugene P. Wigner, among others, seems to think along these lines in his celebrated article 'The Unreasonable Effectiveness of Mathematics in the Natural Sciences', in *Communications on Pure and Applied Mathematics*, XIII, 1960.

2 It was Turing who extended the idea of algorithm as a process, already proposed in 1937, to numerical algorithms and the execution of operations by means of a digital calculator. See A. M. Turing, 'Rounding-off Errors in Matrix Processes' (1948), in *Collected Works of A. M. Turing*, ed. J. L. Britton (Amsterdam: North-Holland, 1992).

1. Abstraction, Existence and Reality

1 See Arthur Schopenhauer, *Preisschrift über die Freiheit des Willens* (1839), in *Die beiden Grundprobleme der Ethik*, in *Zürcher Ausgabe. Werke in zehn Bänden* (Zurich: Diogenes, 1977), vol. VI, p. 96: 'Every *existence* presupposes an *essence*: that is to say, every extant thing must also be *something*, to have a determinate essence. The latter cannot *exist* and yet also be *nothing*, that is to say something such as the *Ens metaphysicum*, something equivalent to a thing that is nothing other than being, divested of any determination or qualities and consequently unable to act in the determinate way that follows from it: as an *essentia* cannot be real without *existentia*, so an *existentia* cannot assume reality without *essentia*.

(Kant demonstrated this with the celebrated example of the 100 thalers).'

2 A set K contained within the set of complex numbers is a *number field* if K contains at least a non-null element and, operating on the elements of K through the four rational operations (addition, subtraction, multiplication and division), we still obtain a K element. What is ultimately at stake is a *closed* domain in relation to the four rational operations.

3 Two domains are *isomorphic*, and hence indistinguishable, if there is a bi-univocal correspondence g from the one to the other that sends the result of an operation between two elements x and y from the first domain A to the result of the analogous operation on the corresponding $g(x)$ and $g(y)$ from the second domain B. If the operation is a sum, we obtain $g(x + y) = g(x) + g(y)$. The notion of 'n-dimensional vectorial space [*sui reali*]' is sufficient to identify a single mathematical object devoid of isomorphism, because a theorem established that any two n-dimensional vectorial spaces [*sui reali*] are isomorphic. A comparable situation pertains to the field of rational numbers Q, with the ordinary operations of addition and multiplication, and the consequent relation of order between fractions ($^n/_m < {}^p/_q$ if $nq < mp$). If a number field K with operations of addition and multiplication and with ordered relations is contained within every other field with addition and multiplication and with ordered relations, then K coincides with Q. See B. Artmann, *Der Zahlbegriff* (Göttingen: Vandenhoeck & Ruprecht, 1983), pp. 18–19.

4 S. Körner, *The Philosophy of Mathematics* (1960) (Dover: New York, 1986), p. 36.

5 W. V. Quine, *From a Logical Point of View* (Cambridge, Mass., Harvard University Press, 1953). See also J. J. Katz, *Realistic Rationalism* (Cambridge, Mass.: The MIT Press, 1998), pp. 117–75.

6 N. Goodman and W. V. Quine, 'Steps toward a Constructive Nominalism', in *Journal of Symbolic Logic*, XII, 1947, p. 105.

7 Valid for all others is the example of complex numbers introduced by Rafael Bombelli in the wake of studies of third-degree equations, a necessary step in establishing the fundamental algebraic theorem with which an algebraic equation always has a solution. More precisely: if $p(x)$ is a polynomial of degree $n > 0$ with coefficients in a field K, then there exists an element z in K such that $p(z) = 0$.

8 This realist definition, owed to Michael Dummett, is related in H. Putnam, *Mathematics, Matter and Method* (Cambridge: Cambridge University Press, 1975), pp. 69–70.

9 The thesis is advanced by Putnam, ibid., pp. 70ff.

10 Ibid., p. 70. For Michael Dummett, Putnam reminds us, in the formulation of a philosophical realism a fundamental ingredient involves the perception of an *external* world beyond our minds.

11 Simone Weil, *Cahiers*, in *Oeuvres complètes*, vol. VI, part III, ed. F. de Lussy (Paris: Gallimard, 2002), p. 179.

12 Bertrand Russell, *Introduction to Mathematical Philosophy* (London: Allen & Unwin, 1919).

2. Mathematics of the Gods

1 G. Scholem, *Über einige Grundbegriffe des Judentums* (Frankfurt: Suhrkamp, 1970), pp. 90–91.

2 The property of closure particularly preoccupied Dedekind, with his recursive theory, in his work *Was sind und was sollen die Zahlen?* (Braunschweig: Vieweg, 1888).

3 See A. Bürk, 'Das Āpastamba-Śulba-Sūtra', in *Zeitschrift der Deutschen Morgenländischen Gesellschaft*, LVI, 1902, p. 336. Equally important are the articles by A. Seidenberg, starting with 'The Ritual Origin of Geometry', in *Archive for History of Exact*

Sciences, I, 1961, where we also find cited the studies by G. Thibault on the *Śulvasūtra* of Baudhāyana, 'The Śulvasútra of Baudhāyana, with the Commentary of Dvárakánáthayaj-van', in *Pandit*, IX, 1874; X, 1875; N.S., I, 1876–77.

4 For a definition of the gnomon see Chapter 5 below, in which the relations between the geometry of Vedic altars and ancient and modern numerical algorithms will be clarified.

5 G. Thibaut, 'On the Śulvasútras', in *Journal of the Asiatic Society of Bengal*, XLIV, 1875, p. 242.

6 This is a crucial passage in the history of mathematics. For a more detailed approach, see D. T. Whiteside, 'Patterns of Mathematical Thought in the Later Seventeenth Century', in *Archive for History of Exact Sciences*, I, 1961, pp. 205–7.

7 See for example G. W. Leibniz, 'Historia et origo calculi differentialis' (1714–16), in *Mathematische Schriften*, vol. V, ed. C. I. Gerhardt (Hildesheim: Olms, 1962).

8 C. B. Boyer, *The History of the Calculus and Its Conceptual Development* (New York: Dover, 1959), p. 118.

9 'The self-existent rendered the external access points [*parāñci khāni*] incapable [of capturing it]: as a result [the individual being] sees only external things and not the inner ātman', Śankara, *Katha Upanishad*, II, 1, 1. On outwardly directed desire (*kāma*), expression of the external condition as opposed to the desire directed towards the *ātman*, see *Taittirīya Upanishad*, with commentary by Śankara (Rome: Āśram Vidyā, 2006), pp. 42, 99 and 114. It is difficult not to be reminded here of Brouwer, the great Dutch mathematician responsible for some of the most important mathematical discoveries of the nineteenth century. Brouwer was the originator of a new mathematics conceived as the fruit of an introspective action – one which did not, however, jeopardize the objectivity of its formulas.

10 For more details see Chapter 4 below.

11 Boethius, *De institutione arithmetica*, ed. G. Friedlein (Leipzig: Teubner, 1867), p. 12: '. . . *eum quoque numerum necesse est in propria semper sese habentem aequaliter substantia permanere, eumque compositum non ex diversis . . . sed ex ipso videtur esse compositus*'.

3. Mathematical and Philosophical Formulas

1 I. Thot, *I paradossi di Zenone nel 'Parmenide' di Platone* (Naples: Bibliopolis, 2006), p. 18.

2 K. Gaiser, *Platons ungeschriebene Lehre* (Stuttgart: Klett, 1963), p. 15.

3 For the use of this expression by Croce, especially with regard to his polemic with Enriques, see L. Russo and E. Santoni, *Ingegni minuti: una storia della scienza in Italia* (Milano: Feltrinelli, 2010).

4 O. Neugebauer, *The Exact Sciences in Antiquity* (New York: Dover, 1969), p. 35.

5 A. P. Youschkevitch, *Les Mathématiques arabes (VIII–XV siècles)*, (Paris: Vrin, 1976), p. 47.

6 The method that we can speculate guided the scribe may be summarized in the iterative formula $x_{k+1} = \frac{x_k + y_k}{2}$ with $x_0 = \frac{3}{2}$ and $y_k = \frac{2}{x_k}$, where k assumes the values 0, 1, 2 . . . The number $\sqrt{2}$ is then larger than y_k and smaller than x_k. That is to say that the interval $[y_k, x_k]$ includes $\sqrt{2}$ for every k and becomes smaller as k grows. The formula is an adaptation of the one Newton and Raphson would use in the seventeenth century. See *Mathematical Cuneiform Texts*, ed. O. Neugebauer, A. Sachs and A. Goetze (New Haven: American Oriental Society and American Schools of Oriental Research, 1945), pp. 42–3; D. H. Fowler and E. Robson, 'Square Root Approximations in Old Babylonian Mathematics: YBC 7289 in Context', in *Historia Mathematica*, XXV, 1998; N. Mackinnon,

'Homage to Babylonia', in *Mathematical Gazette*, LXXVI, 1992.

4. *Growth and Decrease, Number and Nature*

1 Simone Weil felt that every aspect of mathematics could be related to these two principal points, *Cahiers*, in *Oeuvres complètes*, vol. VI, part I, ed. F. de Lussy (Paris: Gallimard, 1944), p. 129.

2 According to E. Bréhier's translation, cited in Aristotle, *La Métaphysique*, ed. J. Tricot (Paris: Vrin, 1970), vol. I, p. 23.

3 Chantraine translated *ousía* as 'reality', 'substance', 'essence'.

4 Martin Heidegger, 'Vom Wesen und Begriff der Phýsis. Aristoteles, Physik B, 1' (1939), in *Gesamtausgabe* (Frankfurt: Klostermann), vol. IX: *Wegmarken*, ed. F.-W. von Herrmann, 1976, p. 272.

5 Ibid., pp. 278–9.

6 K. Kerényi, *Dionysos* (1976) (Stuttgart: Klett-Cotta, 1994), p. 25.

7 Numenius, fr. 2 Des Places.

8 Archimedes, *De sphaera et cylindro*, II, p. 65, rr. 15–16 Mugler.

9 C. B. Boyer, *The History of the Calculus and Its Conceptual Development* (New York: Dover, 1959), pp. 193–4.

10 For such definitions it is possible to consult the unpublished manuscripts in the University of Cambridge Library. In particular, for instance, Newton Papers, ms. Add 3963, f. 47r. For a more extensive discussion see D. T. Whiteside, 'Patterns of Mathematical Thought in the Later Seventeenth Century', in *Archive for History of Exact Sciences*, I, 1961, p. 375.

11 In Platonic thought the soul has this founding principle of the maintenance and cohesion of bodies that would otherwise be subject to incessant change and finally to actual disintegration. See Numenius, fr. 4b Des Places.

5. Katà gnómonos phýsin*: The Nature of the Gnomon*

1 A. Seidenberg, 'The Ritual Origin of Geometry', in *Archive for History of Exact Sciences*, I, 1961, p. 509.

2 Marcel Proust, *Du côté de chez Swann*, in *À la recherche du temps perdu*, ed. J.-Y. Tadié (Paris: Gallimard, vol. I, 1987), p. 182.

3 We are dealing with the so-called Newtonian methods. To calculate the local minimum of a function $f(x)$, where x is a vector of n real components, we augment step by step the x in the direction d in which the function decreases, as one might obtain the valley from the summit of a hill through small movements, choosing with each of them a downhill direction. The iteration is summarized in the formula: $x_{k+1} = x_k + \lambda_k d_k$, assigning an initial value x_0, where λ_k is a number that fixes for every k the length of the passage in the direction d_k. It is shown that if d_k is *any direction downhill* of f, such that f assumes in x_{k+1} an inferior value to the one assumed in x_k, then d_k is the solution of a system of linear equations, the matrix of the coefficients A_k being symmetrical and positive. The method then assumes the form: $x_{k+1} = x_k + A_k^{-1} g_k$, where A_k^{-1} is the inverse of A_k and g_k is the vector of the prime derivatives of f (the so-called gradient) calculated in x_k. The formula generalizes Newton's schema for solving an equation $f(x) = 0$, or rather (for $n = 1$) $x_{k+1} = x_k - \frac{f(x_k)}{f'(x_k)}$, that derives in turn from the technique of gnomonic growth of the square. See C. Di Fiore, S. Fanelli and P. Zellini, 'Low Complexity Secant Quasi-Newton Minimization Algorithms for Nonconvex Functions', in *Journal of Computational and Applied Mathematics*, CCX, 2007, p. 172.

4 The simplest example is that of the square root of a number m. If $m = 2$, Newton's method consists of calculating fractions that approximate $\sqrt{2}$ iteratively, through successive changes, according to the formula $a_{k+1} = \frac{a_k^2+2}{2a_k}$, where the index k assumes the values 0, 1, 2 . . . and one assumes an initial approximation designated a_0, generally included between 1 and 2. If $a_k = \frac{p_k}{q_k}$, then it follows that $a_{k+1} = \frac{p_{k+1}}{q_{k+1}} = \frac{p_k^2+2q_k}{2p_kq_k}$, and hence the numerator and denominator grow quadratically. The number of their digits is doubled, approximately, at every step k.

6. Dýnamis: *The Capacity to Produce*

1 T. L. Heath, *Commentary on the Thirteen Books of Euclid's 'Elements'* (New York: Dover, 1956), vol. I, p. 348.
2 *Mathematical Cuneiform Texts*, ed. O. Neugebauer, A. Sachs and A. Goetze (New Haven: American Oriental Society and American Schools of Oriental Research, 1945), p. 130.
3 J. Høyrup, 'Pythagorean 'Rule' and 'Theorem' – Mirror of the Relation between Babylonian and Greek Mathematics' (Roskilde: Roskilde University, 1999), www.Academia.edu/3131799/.
4 See A. Bürk, 'Das Āpastamba-Śulba-Sūtra', in *Zeitschrift der Deutschen Morgenländischen Gesellschaft*, LV, 1901, p. 556 and LVI, 1902, p. 329; O. Becker, *Das mathematische Denken der Antike* (1957) (Göttingen: Vandenhoeck & Ruprecht, 1966), p. 33.
5 Bürk, 'Das Āpastamba-Śulba-Sūtra', LVI, 1902, p. 327.
6 J. Høyrup, '*Dýnamis*, the Babylonians, and Theaetetus 147c7–148d7, in *Historia Mathematica*, XVII, 1990, p. 208. Høyrup points out that Taisbak interprets *dýnamis* as 'extension'.
7 Bürk, 'Das Āpastamba-Śulba-Sūtra', LVI, 1902, p. 329.

8 Martin Heidegger, 'Die Frage nach der Technik' (1953), in *Gesamtausgabe* (Frankfurt: Klostermann), vol. VII: *Vorträge und Aufsätze*, ed. F.-W. von Herrmann, 2000, pp. 11ff.

9 Iamblichus, *In Nicomachi arithmeticam introductionem*, 11 Teubner.

10 D. J. O'Meara, *Pythagoras Revived* (Oxford: Clarendon Press, 1989), p. 44.

11 Simone Weil, *Cahiers*, in *Oeuvres complètes*, vol. VI, part II, ed. F. de Lussy (Paris: Gallimard, 1997), p. 74.

12 The calculation of the relations defined by diagonal and lateral numbers to approximate $\sqrt{2}$ began from the unit called *spermatikòs lógos* by Theon of Smyrna (first–second century AD).

13 O'Meara, *Pythagoras Revived*, p. 62.

14 M. Psellus, 'Physical Numbers', in *On Pythagoreanism V–VII*, ibid., pp. 218–19.

7. Intermission: Spiritual Mechanics

1 B. Snell, *Die Entdeckung des Geistes* (Hamburg: Claassen & Goverts, 1946).

2 Simone Weil, *Cahiers*, in *Oeuvres complètes*, vol. VI, part IV, ed. F. de Lussy (Paris: Gallimard, 2006), p. 336.

3 J. W. von Goethe, 'Gott und Welt, Urworte. Orphisch', in *Werke. Hamburger Ausgabe in 14 Bänden*, ed. E. Trunz (Munich: dtv, 1998), Vol. I, p. 359.

4 Friedrich Nietzsche, *Al di là del bene e del male* (*Beyond Good and Evil*), in *Opere*, ed. G. Colli and M. Montinari, vol. VI, pt II (Milan: Adelphi, 1968), p. 13.

8. Zeno's Paradoxes: The Explanation of Movement

1 T. L. Heath, *A History of Greek Mathematics*, vol I (New York: Dover, 1981), p. 279; H. Weyl, *Philosophy of Mathematics and*

Natural Science (Princeton: Princeton University Press, 1949), p. 42; A. Grünbaum, 'Can an Infinitude of Operations be Performed in a Finite Time?', in *Philosophical Problems of Space and Time* (Dordrecht: Reidel, 1973).

2 A segment is finite or infinite by division (*katà diaíresin*), whereas a straight line is infinite by its extremities (*toîs eschátois*).

3 Heath, *A History of Greek Mathematics*, vol. I, p. 276.

4 Loc. cit.

5 A. N. Whitehead, *Process and Reality* (1929) (New York: Free Press, 1969), pp. 84–5.

6 F. Cajori, 'The History of Zeno's Arguments on Motion', in *American Mathematical Monthly*, XXII, 1915.

7 Bertrand Russell, *The Principles of Mathematics* (Cambridge: Cambridge University Press, 1903).

8 G. Colli, *Zenone di Elea* (Milan: Adelphi, 1998), p. 121.

9 Quoted in Whitehead, *Process and Reality*, p. 84.

10 Ibid., p. 76.

11 Henri Poincaré, *La Science et l'hypothèse* (1902) (Paris: Flammarion, 1968), p. 51.

12 The number $\sqrt{2}$ is hence associated with a Euclidean construction [*con riga e compasso*]. As is well known, not all the (algebraic) irrational numbers correspond to Euclidean constructions. This circumstance could have an impact on the value of reality [*valore di realtà*] that we decide to attribute to numbers, but this aspect of numerical ontology is beyond our scope here.

13 Whitehead, *Process and Reality*, p. 89.

14 D. Hilbert and P. Bernays, *Grundlagen der Mathematik*, vol. 1 (Berlin: Springer, 1934), pp. 15–17, cited in S. C. Kleene, *Introduction to Metamathematics* (New York: Van Nostrand, 1952), pp. 54–5.

9. The Paradoxes of Plurality

1 Euclid, *Elements*, I, 1: 'A point is that which has no parts.' The term *semeîon*, used by Euclid, seems to assign to the point a degree of reality which is lesser than the Aristotelian *stigmé*, which refers to a kind of signature [*puntamento*]. See T. L. Heath, *The Thirteen Books of Euclid's 'Elements'* (New York: Dover, 1956), vol. I, pp. 155–6.

10. The Limited and the Limitless: Incommensurability and Algorithms

1 J. Stenzel, *Zahl und Gestalt bei Platon und Aristoteles* (Leipzig–Berlin: Teubner, 1924).

2 *The Mānava-Śrautasūtra*, trans. J. M. van Gelder, Śata-Piṭaka Series, vol. XXVII (New Delhi: International Academy of Indian Culture, 1963), p. 308.

3 Ibid., p. 300. The formula is reminiscent of the analogous Platonic formula of 'more or less' that is recurrent in the Dialogues.

4 W. Knorr, 'Aristotle and Incommensurability: Some Further Reflections', in *Archive for History of Exact Sciences*, XXIV, 1981.

5 One obtains $l = d' + l'$ because d', the diagonal DF of Q′, is equal to double the side of the square of diagonal l' ($d' = 2(\frac{l'}{2\sqrt{2}}) = \sqrt{2}\,l'$) and $l' =$ CF = FE for the congruence between triangles BCF and BFE. See Knorr, 'Aristotle and Incommensurability' and H. Rademacher and O. Toeplitz, *Von Zahlen und Figuren* (Berlin: Springer, 1933).

6 A reasoning *ad absurdum* analogous, on the one hand, to that of the *method of exhaustion*, would rely on Archimedes' postulate ('given two homogeneous magnitudes, there always exists a

multiple of the lesser one that exceeds the greater one'), from which Euclid's theorem derives (*Elements*, X, 1): 'Two unequal magnitudes being set out, if from the greater there is subtracted a magnitude greater than its half, and from that which is left a magnitude greater than its half, then there will be left some magnitude less than the lesser magnitude set out.' In the case of the square, if there existed a shared measure m, with *antanaíresis* one would calculate a remainder greater than 0 that is less than m and divisible by m: an absurdity. Archimedes' proposition excludes the existence of infinitesimals that, added an arbitrary number of times, would never exceed a designated line. None of these entities could define a common measure of two magnitudes.

7 We have precisely: $d = l + l' = 2l' + d'$, $l = l' + d'$, and $l' = d - l$, $d' = 2l - d$, where d, l, d' and l' are the diagonal and the side, respectively, of the larger and smaller squares. The vector of the components d and l is obtained by multiplying the matrix M with elements 1 and 2 on the first row and 1 and 1 on the second row for the vector of the components d' and l', while the vector of the components d' and l' is obtained by multiplying the inverse of M by the vector of components d and l. The *geometrical* relations between diagonals and sides suggest a *numerical* progression. Now one may calculate recursively the *numbers* d and l, fixing for each one an initial value equal to 1 and using the preceding relations in the sense of a progressive *growth* of the squares. The first calculated values will be, respectively, 3 and 2. The relation $d{:}l$ then assumes the values 1:1, 3:2, 7:5, 17:12, 41:39, in which the numbers d and l *increase* according to the above-mentioned recursive law.

8 What could be considered as a *relation* $d{:}l$ can also be interpreted as a *fraction* d/l. There is implicit here the use of a unit of measurement equal to $1/l$. The fraction $17/12$ denotes a quantity equal to 17 times the unit of measurement equal to $1/12$. As

the growth of l manifests itself, the unit of measurement $^1/_l$ diminishes, as does the distance between $^d/_l$ and $\sqrt{2}$: another indication of the combination of 'large' and 'small' that designates the nature of the Platonic *ápeiron*.

9 See my *Gnomon* (Milan: Adelphi, 1999), pp. 343–4 and 384–5.

10 Simone Weil highlights this well, with precise theological support: 'Saint John does not assert: we shall be happy because we will see God, but rather: we will be similar to God, because we will see him as he is (1 John, III, 2). We shall be pure good. We will no longer exist. But in this nothing that is at the end of goodness, we shall be more real than in any moment of our earthly life. Whereas the nothingness that is at the limit of evil has no reality. Reality and existence are two different things.' (*Cahiers*, in *Oeuvres complètes*, vol. VI, part IV, ed. F. de Lussy (Paris: Gallimard, 2006), p. 214).

11 One obtains $d^2 = 2l^2 \pm 1$ and hence the relation $d^2{:}l^2$ converges with 2 at the indefinite growth of l.

12 Simone Weil, *Cahiers*, in *Oeuvres complètes*, vol. VI, part III, ed.? (Paris: Gallimard, 2002), p. 139.

13 I. Thomas, *Greek Mathematical Works* (1939) (Cambridge, Mass.: Harvard University Press, 1967), vol. I, pp. 134–5.

14 The machine numbers of the form $(d^1, d^2 \ldots d^t)\,B^r$, where B is the basis of the representation of the numbers. In this case the size of the numbers is represented not by the quantity of the digits but by the exponent r.

15 The *stability* of a procedure of calculation depends on the way in which it propagates on the single arithmetical operations the error due to rounding. The number of such operations, largely inaccessible to the human subject, is usually of a very high order.

16 For example, if a and b are very large numbers, in the first procedure the product $(a \times b)$ could be too large a number for the memory of the calculator and would therefore cause an

overflow. This inconvenience, which would stop the process, could be avoided in the second procedure, in the case in which *c* was a sufficiently small number.

17 A. Bürk, 'Das Āpastamba-Śulba-Sūtra', in *Zeitschrift der Deutschen Morgenländischen Gesellschaft*, LV, 1901, p. 557. There is a surprising affinity with Augustine's thought in the *Soliloquies*, where he explains how from the existence of geometrical figures one may infer the existence of truth and the immortality of the soul, or indeed the intelligence that conceptualizes them. A comparable conclusion is encouraged by Plato's discourses relating to *anámnesis* illustrated in the procedure to double a square (*Meno*, 85 d).

11. The Reality of Numbers: Cantor's Fundamental Sequences

1 Georg Cantor, 'Extension d'un théorème de la théorie des séries trigonométriques', in *Acta Mathematica*, II, 1883, p. 337.

2 B. Artmann, *Der Zahlbegriff* (Göttingen: Vandenhoeck & Ruprecht, 1983), 'Abschlussbemerkungen zu Kapitel 2'.

3 Loc. cit.

4 Georg Cantor, 'Fondements d'une théorie générale des ensembles', in *Acta Mathematica*, II, 1883, p. 390. The italics are mine.

5 P. E. B. Jourdain, 'Introduction' to Georg Cantor, *Contributions to the Founding of the Theory of Transfinite Numbers* (New York: Dover, 1955), p. 67. The italics are mine.

6 J. W. Dauben, *Georg Cantor: His Mathematics and Philosophy of the Infinite* (Cambridge, Mass.: Harvard University Press, 1979), pp. 126–8.

7 Gottlob Frege, *Die Grundlagen der Arithmetik* (Breslau: Koebner, 1884).

8 Immanuel Kant, *Kritik der reinen Vernunft. Zweite Auflage 1787*, in *Werke. Akademieausgabe*, vol. III (Berlin: Reimer, 1904), p. 301.
9 Henri Bergson, *La Pensée et le mouvant*, in *Oeuvres*, ed. A. Robinet (Paris: Presses Universitaires de France, 1959), p. 1254.
10 Ibid., p. 1257.
11 Henri Bergson, *Essai sur les données immédiates de la conscience*, in *Oeuvres*, p. 72.
12 N. Wiener, *Extrapolation, Interpolation, and Smoothing of Stationary Time Series* (1942: first version with limited circulation; and 1949: more extended version, including the articles by N. Levinson that appeared in 1943 in periodical form) (Cambridge, Mass.: MIT Press, 1949). The Toeplitz matrices feature in an article by Levinson that simplifies and clarifies many aspects of Wiener's original work: 'The Wiener RMS (Root Mean Square) Error Criterion in Filter Design and Prediction', in *Journal of Mathematics and Physics*, XXV, 1946, reprinted as Appendix B in Wiener, *Extrapolation, Interpolation, and Smoothing of Stationary Time Series*.
13 The inverse of a matrix A of n rows and n columns is a matrix B such that the product of B times A is equal to the identity matrix I, which has 1 on the principal diagonal and 0 elsewhere. Hence we write $B = A^{-1}$. If A is the matrix of coefficients of a system of linear equations, the knowledge of its inverse A^{-1} allows us to easily calculate the solution of the system as a simple product of A^{-1} for the vector of known terms. See below, note 1 to Chapter 16.

12. The Reality of Numbers: Dedekind's Sections

1 See R. Smith, 'The Mathematical Origins of Aristotle's Syllogistic', in *Archive for History of Exact Sciences*, XIX, 1978.
2 H. Reichenbach, *Philosophie der Raum-Zeit-Lehre* (Berlin–Leipzig: de Gruyter, 1928), par. 9. Reichenbach argues that the normative value of Euclidean mathematics resides in the logic of

demonstration much more than in the visualizable construction of the figures. It is possible to think by way of analogy of the much discussed thesis that the mathematics of Euclid's *Elements* is a geometrical algebra.

3 See for example R. Courant and H. Robbins, *What is Mathematics?* (London: Oxford University Press, 1941).

4 Richard Dedekind was the first to conceive of the theory of real numbers, indicating as decisive (in 1858) the property of *completeness.* His fundamental treatise, *Continuity and Irrational Numbers* (1872), summarizes in brief his essential ideas.

5 See T. L. Heath, *Commentary on The Thirteen Books of Euclid's 'Elements'* (New York: Dover, 1956), vol. II, p. 125.

6 To say that the relation $a{:}b$ between two *magnitudes* a and b is less than the relation $m{:}n$ between two whole *numbers* m and n means that $na < mb$. The same applies to the relations of 'equal to' and 'greater than'.

7 For a comprehensive discussion of *antanaíresis* in pre-Euclidean mathematics, see in particular D. H. Fowler, *The Mathematics of Plato's Academy* (Oxford: Clarendon Press, 1990).

8 B. Riemann, *Über die Hypothesen, welche der Geometrie zu Grunde liegen* (Göttingen: Dieterichsche Buchhandlung, 1867).

9 See above, note 6.

10 See G. H. Hardy and E. M. Wright, *An Introduction to the Theory of Numbers* (1938), fifth edition (Oxford: Oxford University Press, 1979), pp. 138–9.

11 Simone Weil, *Oeuvres complètes*, vol. VII, part I, ed. F. de Lussy (Paris: Gallimard, 2012), p. 465. The italics are mine. On this subject see Roberto Calasso, *Il Cacciatore Celeste* (Milan: Adelphi, 2016), p. 272.

12 A. M. Turing, 'On Computable Numbers, with an Application to the *Entscheidungsproblem*', in *Proceedings of the London*

Mathematical Society, second series, XLII, 1937; 'A Correction', ibid., second series, XLIII, 1938.

13 W. V. Quine, *Set Theory and Its Logic* (Cambridge, Mass.: Harvard University Press, 1963).

14 See also S. C. Kleene, *Mathematical Logic* (New York: Wiley, 1967), Ch. 2.

15 K. Scheel, *Der Briefwechsel Richard Dedekind–Heinrich Weber*, ed. T. Sonar and K. Reich (Berlin: de Gruyter, 2014), p. 277. The emphasis added is mine; B. Artmann, *Der Zahlbegriff* (Göttingen: Vandenhoeck & Ruprecht, 1983), p. 65.

16 There being an isomorphism between the set of algebraic numbers (roots of an algebraic equation, that is to say, of a polynomial equal to 0) and the system of decimal numbers of the form $a \times 10^{-k}$, with a an integer and k a variable between natural numbers, the set of algebraic numbers also has some lacunae: ibid., p. 42.

17 Georg Cantor, 'Extension d'un théorème de la théorie des séries trigonométriques', in *Acta Mathematica*, II, 1883, p. 340.

18 Georg Cantor, 'Fondements d'une théorie générale des ensembles', in ibid., p. 393.

19 Ibid., p. 403.

20 The Archimedean property, already identified by Greek mathematicians, obtains when, given x and y, with x smaller than y, there is a multiple of x that is superior to y. What is demonstrated is that an ordered and complete numerical field satisfies this property.

21 Artmann, *Der Zahlbegriff*, p. 33.

22 U. Dini, *Fondamenti per la teorica delle funzioni di variabili reali* (Pisa: Nistri, 1878), p. 4.

23 Ibid., p. 6.

24 A. N. Whitehead and Bertrand Russell, *Principia mathematica* (Cambridge: Cambridge University Press, 1973), pp. 71–2.

25 Quine, *Set Theory and Its Logic*, p. 3.
26 J. W. Dauben, *Georg Cantor: His Mathematics and Philosophy of the Infinite* (Cambridge, Mass.: Harvard University Press, 1979), pp. 221–2.
27 Friedrich Nietzsche, *Frammenti postumi, 1887–1888*, in *Opere*, ed. G. Colli and M. Montinari, vol. VIII, pt II (Milan: Adelphi, 1974), p. 47.

13. Mathematics: A Discovery or an Invention?

1 Richard Dedekind, *Was sind und was sollen die Zahlen?* (Braunschweig: Vieweg, 1888), p. III.
2 Ibid., p. VI. The italics are mine.
3 Richard Dedekind, *Stetigkeit un irrationale Zahlen* (Braunschweig: Vieweg, 1872), pp. 18 and 20–22.
4 S. Pincherle, 'Saggio di una introduzione alla teoria delle funzioni analitiche secondo i principii del Prof. C. Weierstrass', in *Giornale de Matematiche*, XVIII, 1880, pp. 186, 190 and 191.
5 Cited in F. Cajori, *A History of Mathematical Notations* (1928–9) (New York: Dover, 1993), vol. II, p. 333. One is prompted to make a comparison with Benedetto Croce's exhortation to philosophers: 'Think, and don't calculate! [*Qui incipit numerare, incipit errare!*]': Benedetto Croce, 'Filosofia della pratica. Economia ed etica', in *Filosofia dello spirito*, vol. III (Bari: Laterza, 1909, second edition revised by the author, 1915), p. 270. Implicit within this is an observation analogous to one made by Whitehead but turned into a negative.
6 G. H. Hardy, *A Mathematician's Apology* (Cambridge: Cambridge University Press, 1940).

14. From the Continuum to the Digital

1 Richard Dedekind, *Stetigkeit und irrationale Zahlen*, (Braunschweig: Vieweg, 1872), pp. 18–19.

2 'Das Kontinuum als Medium freien Werdens' is Hermann Weyl's expression. See H. Weyl, 'Über die neue Grundlagenkrise der Mathematik', in *Mathematische Zeitschrift*, X, 1921.

3 Friedrich Nietzsche, *Frammenti postumi, 1888–1889*, in *Opere*, ed. G. Colli and M. Montinari, vol. VIII, pt II (Milan: Adelphi, 1974), p. 162.

4 See D. Hilbert, 'Über das Unendliche', in *Mathematische Annalen*, XCV, 1926.

5 É. Borel, 'Les "Paradoxes" de la théorie des ensembles', in *Annales scientifiques de l'École Normale Supérieure*, third series, XXV, 1908.

6 See for example V. A. Uspensky, 'Kolmogorov and Mathematical Logic', in *Journal of Symbolic Logic*, LVII, 1992, p. 393.

7 See E. Zermelo, 'Neuer Beweis für die Möglichkeit einer Wohlordnung', in *Mathematische Annalen*, LXV, 1908. For the theory of von Neumann, see P. R. Halmos, *Naive Set Theory* (New York: Springer, 1974).

8 This is the case, for example, of the diagonal numbers d and the lateral ones l. One may calculate k in such a way that the distance between 2 and the relation $d^2{:}l^2$ is less than an ε that is arbitrarily small if l is greater than k, because we know that the distance is equal to $\frac{d^2}{l^2} - 2 = \pm\frac{1}{l^2}$. This implies that the sequence of relations $\frac{d^2}{l^2}$ is *fundamental* in the sense given by Cantor. Nevertheless, as Brouwer demonstrated, it is not always possible to attain an *efficient* calculation for the index k for which the distance between a_{n+p} and a_n is less than ε if $n > k$.

9 G. Peano, 'Interpolazione nelle tavole numeriche', in *Atti della Reale Accademia delle Scienza di Torino*, LIII, 1917–18, p. 693.

10 See R. Courant and D. Hilbert, *Methods of Mathematical Physics*, vol. II: *Partial Differential Equations of R. Courant* (New York: Interscience, 1962), p. 229.

11 Ibid., p. 230.

12 B. Parlett. 'Progress in Numerical Analysis', in *SIAM Review*, XX, 1978, pp. 448–9.

13 A. A. Markov Jr, 'On Constructive Functions', in *American Mathematical Society Translations*, second series, XXIX, 1963, pp. 163–4.

14 A. A. Markov Jr, 'The Theory of Algorithms', in ibid., second series, XV, 1960, p. 3.

15 H. Rogers Jr, *Theory of Recursive Functions and Effective Computability* (New York: McGraw-Hill, 1967), p. 31.

15. The Growth of Numbers

1 For greater detail, see R. Courant and H. Robbins, *What is Mathematics?* (London: Oxford University Press, 1941), pp. 176–80.

2 See M. S. Paterson, 'Efficient Iterations for Algebraic Numbers', in R. E. Miller and J. Thatcher (eds.), *Complexity of Computer Computations* (New York–London: Plenum Press, 1972). In an iterative method of order p which calculates at every step k a rational expression g that provides an approximate value p/q of a root z of an algebraic equation (the *order* measures the speed with which it approximates z), the denominator q, for a quite magnified k, exceeds the value Ld^s with $s = r^k$ for every $r < p$, where L and d are constants, with d greater than 1.

3 For example, thanks to automatic processes similar to rounding, the fraction $1300/1113$ may be approximated by the simpler

fraction 7/6. It would be excessive to simplify a fraction by dividing the numerator and denominator by the same factors. Besides, the probability that two random whole numbers have a prime between them is quite high, equal to $\frac{6}{\pi^2} \approx .60793$. See D. Knuth, *The Art of Computer Programming*, vol. II, second edition (Reading, Mass.: Addison-Wesley, 1981), pp. 315 and 324. See also P. Henrici, 'A Subroutine for Computations with Rational Numbers', in *Journal of the Association for Computing Machinery*, III, 1956, pp. 6–9.

4 D. Knuth, *The Art of Computer Programming*, vol. II (Reading, Mass.: Addison-Wesley, 1969), p. 292.

5 Consider a single algebraic equation, defined by a polynomial in x equal to 0. If the polynomial has a degree greater than 4, the equation does not admit, in general, a solution of an analytical kind. One must therefore approximate the solutions with numerical procedures which calculate sequences of rational numbers that approximate those solutions up to the requisite degree of precision. A comparable situation can be found for differential or integral equations, the solution of which is not expressible with an analytical formula, and where it therefore becomes necessary to approximate it numerically. To this end it is necessary to resolve systems of linear equations with matrices of coefficients of a high dimension.

6 See W. Gautschi, 'Computational Aspects of Three-term Recurrence Relations', in *SIAM Review*, IX, 1967. For example, a recurrence relation with three terms assuming the form $y_{k+1} = ay_k = by_{k-1}$, which makes it possible to calculate y_k, for $k = 2, 3, \ldots$, if they are assigned the initial values y_0 and y_1, is often characterized by phenomena of instability, that is to say, by the uncontrolled growth of error. With comparable iterative strategies, in good order, simple systems of differential equations are solved, and in such cases the uncertainties regarding

the initial values may cause exponential growth of the terms from the oscillating progression that is provoked by such unstable numerical phenomena. On this subject see G. Dahlquist, '33 Years of Numerical Instability, Part I', in *Bit Numerical Mathematics*, XXV, 1985, and G. Dahlquist and Åke Björck, *Numerical Methods* (1974) (New York: Dover, 2003), pp. 342 and 373.

7 Quoted in A. Knoebel, R. Laubenbacher, J. Lodder and D. Pengelley, *Mathematical Masterpieces* (New York: Springer, 2007), p. 65.

16. The Growth of Matrices

1 The vector w has the element of position i as the sum, as k varies, of the products $a_{ik} \times v_k$, where v_k is the element of position k of the vector v. Analogously, the product of two square matrices of n rows and n columns A and B is the square matrix C, of the same dimensions as A and B, the element c_{ij} of which is the sum, when k varies, of all the products $a_{ik} \times b_{kj}$. If the product of the two matrices A and B is the identity matrix I, then B is the inverse of A, and A is called *invertible* and we write $B = A^{-1}$. A vector solution x of a system of linear equations $Ax = b$ with matrix A having invertible coefficients is equal to the product of A^{-1} for the vector b. See note 13 to Chapter 11 above.

2 V. Y. Pan, 'Complexity of Parallel Matrix Computations', in *Theoretical Computer Science*, LIV, 1987.

3 G. Strang, 'Wavelets', in *American Scientist*, LXXXII, 1994.

4 See J. H. Wilkinson, 'Some Comments from a Numerical Analyst' (1970 Turing Lecture), in *Journal of the Association for Computing Machinery*, XVIII, 1971. On the initial stages of the development of matrix calculation and on the role of Whittaker and Robinson's book, see O. Taussky, 'How I Became a Torchbearer for Matrix Theory', in *American Mathematical Monthly*,

XCV, 1988. For the first signs of the inefficiency of Cramer's method, see G. E. Forsythe, 'Solving Linear Algebraic Equations Can be Interesting', in *Bulletin of the American Mathematical Society*, LIX, 1953.

5 D. Hilbert, 'Ein Beitrag zur Theorie des Legendre'schen Polynoms', in *Acta Mathematica*, XVIII, 1894.

6 The expression of the error contains the powers of the variable x which, when x is between 0 and 1, are close to being linearly dependent. For this reason the strategy for reducing the error to a minimum returns to a system of linear equations identified by a matrix H the rows of which are close to being linearly dependent, with the consequence that the determinant of H is close to 0, which is a sign of the ill-conditioned nature of H. In fact, the reciprocal of the condition index of a matrix H is of the set of matrices with null determinant. Consequently, the conditioning index is bigger the smaller this distance turns out to be. See J. W. Demmel, 'On Condition Numbers and the Distance to the Nearest Ill-posed Problem', in *Numerische Mathematik*, LI, 1987.

7 The matrix is described in its previous explicit form in G. E. Forsythe, 'Pitfalls in Computation, or Why a Math Book isn't Enough', in *American Mathematical Monthly*, LXXVII, 1970. See also G. E. Forsythe and C. B. Moler, *Computer Solution of Linear Algebraic Systems* (Englewood Cliffs, NJ: Prentice-Hall, 1967).

8 See note 1 above for the expressions that define the product of a matrix for a vector. The solution of the system of six equations is obtained by multiplying the matrix H^{-1} of six rows and six columns for the vector b, obtaining thereby a vector solution x of which the fifth element x_5 is given by an expression containing the term $4410000b_5$, where b_5 is the fifth element of the vector b. If b_5 is subject to a small variation, a positive increment, let us say, equal to 10^{-6}, instead of $4410000b_5$

we would have $4410000(b_5 + 10^{-6})$; that is to say, x_5 is subject, due only to the disturbance 10^{-6}, to a variation equal to 4.41. An equivalent amplification of the error may be verified, naturally, for any sufficiently high element of the inverse of H.

9 For greater detail, see N. J. Higham, *Accuracy and Stability of Numerical Algorithms* (Philadelphia: SIAM, 1996), p. 177.

10 A real symmetric matrix A (one, that is, for which $a_{ij} = a_{ji}$) is *positive definite* if $x^T A x > 0$ for every non-null real vector x. The matrix A is positive definite if and only if its eigenvalues are (real and) positive.

11 More precisely, if X is the *calculated* inverse of A, the distance between AX and the identity matrix I will not be greater than the value $14.24\left(\frac{\lambda_{max}}{\lambda_{min}}\right)n^2\beta^{-s}$, where λ_{max} and λ_{min} are, respectively, the maximum and minimum eigenvalues of A, and the calculation takes place with a floating point with a number of significant figures s and a representation base equal to β.

12 J. H. Wilkinson, 'Modern Error Analysis', in *SIAM Review*, XIII, 1971, p. 550. Wilkinson also underlines the valuable contribution to the analysis of error propounded by Wallace Givens, who was the first to conceive, in 1954, a backwards analysis of the error, consisting of *reflecting* on the matrix the errors accumulated during the calculation: the approximate algorithm is conceived as an algorithm that inverts exactly not the matrix A but a matrix $A + \Delta A$, that is, a matrix perturbed by an error ΔA.

13 A. M. Turing, 'Rounding-off Errors in Matrix Processes' (1948), in *Collected Works of A. M. Turing*, ed. J. L. Britton (Amsterdam: North-Holland, 1992).

14 Frequently the efficiency of known iterative methods, especially the speed of convergence of a solution, depends on the condition index μ.

15 The *Euclidean norm* of A, also known as the *spectral norm*, because if A is symmetrical and positive definite together with

its spectral radius [*raggio spettrale*], that is, with its maximum eigenvalue, it is less than or equal to the Frobenius norm.

16 It *may* be, but is not necessarily so. The error is by itself unknowable; we can only establish its limits. But if these limits are compatible with a large error, every certainty regarding the reliability of the computational method is undermined. There are still methods of *conditioning*, techniques of intervention in the matrix to modify the distribution of the eigenvalues. There is a vast literature on this subject.

17 See above, note 11 of Chapter 2.

18 A. A. Markov Jr, 'The Theory of Algorithms', in *American Mathematical Society Translations*, second series, XXIX, 1963, p. 1.

19 H. H. Goldstine and J. von Neumann, *On the Principles of Large-scale Computing Machines*, in John von Neumann, *Collected Works*, vol. V, ed. A. H. Taub (Oxford: Pergamon Press, 1963), p. 3.

17. The Crisis of Fundamentals and the Growth of Complexity: Reality and Efficiency

1 The determinant of a matrix A is the polynomial function of the elements of A . . .

2 This potential inconvenience was pointed out, perhaps for the first time, by J. Edmonds, 'Systems of Distinct Representatives and Linear Algebra', in *Journal of Research of the National Bureau of Standards*, LXXI B, 1967. See also L. Lovász, 'Algorithmic Aspects of Some Notions in Classical Mathematics', in J. W. de Bakker, M. Hazenwinkel and J. K. Lenstra (eds), *Mathematics and Computer Science* (Amsterdam: North-Holland, 1986).

3 Cf. J. Hartmanis, 'Gödel, von Neumann and the P = ?NP Problem', in *The Structural Complexity Column*, 1989, EATCS Bulletin, vol. 38, pp. 101–7. Hartmanis refers precisely to three fundamental problems – incompleteness, undecidability and

complexity – thus explaining how the last inherits the fundamental question of what, in mathematics, may or may not admit of a solution.

4 Even the calculation of a simple quadratic polynomial $p(x)$ is not free of challenges. If we write the formula of the simple parabola of equation $y = p(x) = 1 + (x - 5555.5)^2$ in the form $p(x) = 30863581.25 - 11111x + x^2$, calculating $p(5555)$ and $p(5554.5)$ in machine arithmetic with six significant digits we obtain, respectively, the numbers 0 and 100. In other words, a small perturbation on x may be reflected in a large perturbation of $p(x)$. The example is taken from J. R. Rice, *Numerical Methods, Software, and Analysis* (New York: McGraw-Hill, 1983), p. 60.

5 A. L. Cauchy, *Cours d'Analyse de l'École Royale Polytechnique. Première Partie. Analyse Algébrique* (Paris: Debure, 1821), p. 463. The italics are mine.

6 If x belongs to the interval $[c, d]$, then $h(x)$ has an absolute value less than or equal to ε. If in the same interval $t(x)$ is positive, we cannot infer that $h(x)$ follows suit, because the inequality $h(x) + e(x) > 0$ is compatible with the inequality $h(x) < 0$. It is different if x does not belong to $[c, d]$. In that case, $h(x)$ has an absolute value greater than ε, and if $t(x)$ is positive, then $h(x)$ is also positive. In fact, if $h(x) < 0$, it should be the case that $h(x) < -\varepsilon$, and thus also that $h(x) + e(x) < 0$, a contradiction.

18. Verum et Factum

1 Simone Weil, *Cahiers*, in *Oeuvres complètes*, vol. VI, part III, ed. F. de Lussy (Paris: Gallimard, 2002), p. 90.

2 E. P. Wigner, 'The Unreasonable Effectiveness of Mathematics in the Natural Sciences', in *Communications on Pure and Applied Mathematics*, XIII, 1960.

3 A. Capelli, 'Saggio sulla introduzione dei numeri irrazionali col metodo delle classi contigue', in *Giornale di Matematiche*, XXXV, 1897, p. 210.

4 Weil, *Cahiers*, vol. VI, part II, p. 352.

5 Thomas Aquinas, *Summa Theologiae*, I, q. 4, a. 1: '*Oportet enim ante id quod est in potentia, esse aliquid actu, cum ens in potentia non reducatur in actum, nisi per aliquod ens in actu.*'

19. Recursion and Invariability

1 R. J. Gillings, *Mathematics in the Time of the Pharaohs* (1972) (New York: Dover, 1982). See also S. Couchoud, *Mathématiques égyptiennes* (Paris: Le Léopard d'Or, 1993).

2 Manuscript letter, private collection of the author, 1972.

Index

abstract objects 14
abstraction 11–12, 15, 155, 155–6
actuality 117
Aeschylus 18, 24
Akiva, Rabbi 21
algebra
  geometrical 112
  linear 168
  of matrices 167–8
algebraic automatism 31
algebraic closure 177
algebraic equations 30
  third-degree 22
algebraic numbers 159–60
algorithm procedure 104–13
algorithmic constructions 8, 190
algorithms 4–6, 7–8, 97, 147, 154
  computational efficiency 5–6
  constructability 190
  definition 5
  dichotomous divisions 196–7
  effectiveness 178
  efficiency 178–9, 186–91, 193–4
  and reality 154–6
  for solving equations 39–42
  stability 190, 213n15
  trends 5
analysis, arithmetization of 153–4
analytical formula 151–2
anatanaíresis 129–30, 162
ancient arithmetic 4
Antaeus 155
antanaíresis 37, 107, 111
antinomy, the 120
Āpastamba 27, 29, 62, 113
approximate calculus 105–6
approximation 98, 99, 157, 165
  Hilbert's procedure of 169–72
  irrational numbers 99–113, 158–60, 162–3
  limit 111
  limit of 187–8
  rational 100
Archimedes 40, 90, 91, 157
Archimedes' postulate 211–12n6
Archytas 67

Aristotle 37, 43, 45, 46–8, 51, 67, 72, 73, 74, 78–80, 86, 88, 89–90, 90, 93, 95–6, 100, 112, 119, 125, 129
arithmetic
  ancient 4
  Mesopotamian 6
arithmetization 153–4, 178–9
arithmetizing analysis 22
Armament Research Department 169
astronomical signs 24
atomic propositions 13
atomism 145–6
atomistic realism 117, 119
atomistic theories 91, 94–5, 112
Augustine, St 91, 134, 214n17
automatic calculation 100
automatic learning 60
axioms 12

Babylon 4, 61–2, 63
Baudhāyana 27
becoming 44–5
Being 114
Bergson, Henri 121–2, 124
Berkeley, Bishop 80
Bernays, P. 87
Bernoulli, Jakob 163–4
Bernoulli numbers 163–4
Bessel functions 163–4
binary numbers 160–1
binomial, square of 31
Boethius 35, 177
Bolzano, Bernhard 121, 189
Bombelli, Rafael 22, 30, 203n7
Borel, Émile 5, 8, 147
Brhadāanyaka Upanishad 33
Brouwer, L. E. J. 8, 54, 124, 149
Bruno, Giordano 26
Bürk, Albert 113

calculability 8
  limits of 191
calculation 23–4
  and knowledge 71
  processes 5
  size of 197
calculators 5, 15, 164
calculus 4, 5, 15, 23, 35, 53, 153, 168–9, 180–1
  approximate 105–6
  complexity 184, 185
  differential 31, 58
  digital 194
  Mesopotamian 41
  recursive 196–7
  sensitivity 172
  theoretical bases 150
Cantor, Georg 19, 22, 84, 86, 92–3, 96, 98, 114–24, 126, 132, 133, 134, 137, 146, 149–50, 159–60, 177
Cardano, Girolamo 22
cardinality 92
Cauchy, Augustin-Louis 81, 105, 106, 187

Cavalieri, Bonaventura 31
change 44–5, 48–9, 50–1, 54, 58–9
Chinese mathematics 23
Chrysippus 73
Church, A. 185
class, closed 22
coincidence 66
Colossus of Rhodes 50–1
completeness 127, 132–3, 192, 193
complex numbers 22, 119, 203n7
complexity 183–6
computability 5
computatio 23
computation
  automatic 157–8
  numerical 162–3
  using rational numbers 161
computational complexity 5, 183–6, 198
computational cost 156
computational efficiency 5–6
computational mathematics 150, 178, 186, 191
computational processes 8–9, 153, 187
computational schemas 6
computational strategies 8, 179
condition index 174–6, 224n14
consciousness 53, 54, 68
constructivist philosophies 8
continuity 192
continuity, law of 41
continuous extension, atomized 84
continuous fraction, the 129, 163
continuous mathematics 4
continuum, the 7, 18, 114
  atomism 145–6
  Cantor's model 116–17
  critique of 147–8
  division of 76–88, 95
  as free becoming 144
  infinity of 130–1
  metric theory of 93
  structure of 111
Cramer's algorithm 169, 172–3, 177, 183, 196
Croce, Benedetto 39
curve–straight line relation 43
cybernetics 122

decimals 151
decrease 43, 44, 45, 49–50
Dedekind, Richard 15, 22, 64, 75, 84, 86, 96, 98, 107, 116–17, 118, 120–1, 125–38, 139–40, 141, 144, 146, 148–9, 177–8, 194–5, 216n4
deductions 13
Democritus 94, 95
demonstration procedure, the 102–4
density 128, 133

derivation 22
Descartes, Rene 7, 138
descriptions, theory of 136
deterministic chaos 161
diagonal argument, the 116–17
diagonal method, the 19
diaíresis 110
dialectic 37
dichotomous divisions 196–7
dichotomous method, the 100–2
differential calculus 31, 58
digital calculation 160–1, 162–3, 168
digital calculus 194
digital computation 165–6
digital revolution 153–4
dimensions 92, 166
Dini, Ulisse 134–5, 137–8
discovery 15–16
divine, the 27
division 196–7
  to the extremities 77–80
dynamic systems, theory of 161
dynamics 39
dýnamis 62–3, 65, 90, 181

effective realization 116
efficiency 5–6, 178–9, 183, 186–91, 193–4, 196–7, 224n14
Egypt 196–7
Elea 24
Eleatic logic 114
emboîtement 122–3
enigmas 26
enlargement 28, 28–9, 50–1, 61
Ennead, the, genealogical tree of 197
enumeration 18–19
Epicharmus 44–5, 74–5
equations
  algorithms for solving 39–42
  root of 59–60
  solving 59
equivalence 27–8, 94, 110
Eratosthenes 26–7, 50–1
essentia 11–12
Euclid 15, 18, 28, 40, 61, 62, 90, 100, 112, 125–6, 127, 130, 131, 162, 211n1
Euclidean algorithm, the 37, 102–4, 111
Euclidean norm 176, 224–5n15
Euclid's theorem 57–9, 212n6
Eudoxos/Euclid theory of relations 128–9, 141
Euler, Leonhard 58, 163, 164
Eutocius of Ascalon 26–7
exhaustion, method of 40
existence 108
existentia 11–12
existential quantifier, the 18

false-position method 39–40
feedback 54
Fibonacci 198

fire altars 27–35, **36**, 43, 56, 64, 99
fluid dynamics 180
fluxions 52–3
Forsythe, George 170–1
Foucault, Michel 20
foundational themes 21–2
fractions 36, 111, 115, 158, 163
Frege, Gottlob 12–13, 119–20, 125, 137, 146–7, 177
Frobenius norm 174–5, 176
fundamental sequences 98, 114–24, 126, 133, 192, 193
fundamentals, crisis of 5, 184

Galileo 16, 38–9, 41
Gauss, Carl Friedrich 172–3
Gauss's algorithm 172–3, 183
geometric figures, growth of 7
geometric forms, enlargement of 6–7, 8, 50–1, 61
geometrical algebra 112
geometrical point, the 51–2
geometry 4, 6–7
  Greek 23, 28
  Vedic 23, 27–35, 36, 43, 56, 64, 99, 112–13
Givens, Wallace 224n12
gnomon, the 30, 55–60
gnomonic incrementation 30
Gödel, Kurt 184–6
Gods 6–7, 21
Goethe, Johan Wolfgang 74–5
Goldstine, Herman 173–4, 176, 180, 191
Goodman, N. 14
Gorgias 25
Greece, ancient 6, 8, 17–18, 26–7, 41–2, 61, 98, 157, 181
  geometry 23, 28
  and growth 45–6, 50–2
  and nature 46–9
  and Vedic geometry 28
growth 43–54, 60, 196–200
  and change 44–5, 50–1
  and efficiency 196–7
  of forms 50–1
  and nature 46–9, 51, 90
  and numbers 67
  of numbers 97, 157–64, 197–8
  of the soul 73–4
  speed of 100
  unlimited 45

Hadamard, Jacques 152
Hardy, Godfrey H. 142
Hartmanis, Juris 185
Hegel, G. W. F. 37
Heidegger, M. 47, 64–5, 75
Heliopolis 197
Heraclitus 72, 72–3
Hermes 24
hermetic power, of formulas 38
Hermite, Charles 17, 134
Herodotus 18
Heron of Alexandria 55

Hesiod 18, 25–6
Hilbert, David 87, 146,
    169–72, 184, 187
Hilbert matrix 170–2, 177
Homer 18, 25, 49, 67, 74
Hopf, Eberhard 122
Hotelling, Harold 173

Iamblichus 25, 44, 67, 109
ignorance 3
ill-posed problems 152
immutability 35
incommensurability 4, 130
  demonstrations of 100
  and Vedic geometry 112–13
  visible 111
incommensurable quantities,
    relationships between 41–2
indetermination, cloud of 190
indices 115
indivisibles 31
infinite, the 17–18, 43, 45,
    69, 76
  Cantor's thesis of 118–21
  realizable 118
  Zeno's Paradoxes and 77–80
infinitesimals 80, 81
infinity 17–18, 108
  of the continuum 130–1
  Leibniz 31
information technology
    154, 199
information theory 5
interval of uncertainty 187
intervals, indefinite sequence
    of 106–7
intuitionism 147–50
invariability 45, 198–9
invention 15–16
irrational numbers 7–8, 19, 36,
    82–3, 96, 119, 126–7,
    128–30, 144–5, 210n12
  approximation 99–113,
    158–60, 162–3
  as an indefinite sequence of
    intervals 106–7
isomorphism 12, 117–18,
    202n3, 217n16
iteration 198–9

James, William 83, 138

Kant, I. 17, 120
Katha Upanishad 34
Kātyāyana 27, 29
Kleene, Stephen 87–8, 150
knowledge
  and calculation 71
  demand for new 150–1
  sources of 15
Knuth, Donald 161–2
Kronecker, Leopold 137,
    144–5

Lagrange, Joseph-Louis 58, 151
learning 60
Lebesgue, Henri 144–5
Leibniz, W. 80, 120, 153

Leucippus 94, 95
likeness 129
limit 53, 104–5, 188
approximation 111
concept of 22
definition 99
limits, reaching 80
linear algebra 168
linear equations 165–6, 168, 169, 170, 174–5, 180, 183
linear problems 166
linearization 165–6
lines 92, 114
Liouville, Joseph 100, 158–60
Locke, John 117
logic 12–14, 18, 125–6, 131, 135–7, 137–8
logicism 12–13
lógos 36, 47, 66, 72–3, 97
Lucretius 94–5

Mach, Ernst 15
magnitude 75
Malamoud, Charles 25
Mark, Gospel of 74
Markov, Andrej A., Jr 155, 178
materiality 46
mathematical abstraction 9
mathematical analysis 42
mathematical concepts, behaviour 3
mathematical entities 11–12, 16, 17, 191
essence of 117–18
existence of 147
reality of 193–4
mathematical fields 12
mathematical intuitionism 8, 147–50
mathematical modelling 151–5
mathematical objects 12
mathematical problems, solution 192–4
mathematical reality 5
mathematical thinking 3, 17
mathematics 11–12, 22
aims 117, 151, 179
applications of 16
crucial moments 153–5
discovery or invention debate 139–43
effectiveness 194
efficacy 179–80
forerunner 139
kinds of 20
and nature 35
origins 3–4, 20–2, 22, 23–4, 139–43
perceptions of 3
reality of 142
reductionism 181
and religion 24–6, 26–7
traditions 23
matrices and matrix calculation 215n12, 215n13
algebra of 167–8
condition index 174–6
condition number 175

matrices and matrix
calculation – *cont'd.*
elements 166
error analysis 172–6, 223n6, 224n12, 225n16
growth of 168–82
the Hilbert matrix 170–2, 177
ill conditioned 171
norm 174–5, 176
overflow 176
structure of 166–7, 191
theory of 199–200
vectors 166–8, 222n1
Melissus of Samos 76–7
mental framework 17
Mesopotamian mathematics 6, 23, 39, 41–2, 61, 61–2
metamorphosis 48–9
metaphysics 13, 41, 43
models and modelling 3, 151–5
Moses and Rabbi Akiva, fable of 21
motion, laws of 78
movement 73–4
Zeno's Paradoxes 76–88
mutability 45
mysticism 7, 13

National Physical Laboratory 169
nature 8, 69, 181–2
and growth 46–9, 51, 90
and mathematics 35
and mysticism 7
and numbers 69–70
and the soul 73
Neo-Platonists 69
Neugebauer, Otto 41, 42
neural networks 60
Newton, I. 25, 30, 52–3, 59–60, 70, 80, 153, 158, 205n6
Newtonian methods 207n3
Nicholas of Cusa 36–7
Nicomachus of Gerasa 70
Nietzsche, F. 20, 75, 137–8, 145–6
nominalism 13–14
non-existence 108
nothing 90
NP-complete problems 183–4, 185–6
numbers 11, 216n4
actuality of 118
algebraic 159–60
binary 160–1
completeness 192
complex 22, 119, 203n7
continuity 192
enumeration 19
essence of 12
geometrical form 23
and growth 67
growth of 6, 7, 97, 157–64, 197–8
irrational 7–8, 19, 36, 82–3, 96, 99–113, 106–7, 119, 126–7, 128–30, 144–5, 158–60, 162–3, 210n12

and logic 13, 14
and nature 69–70
ontological status 18, 120, 177
origins 141–2
Pythagorean 55–7, **56**
rational 22, 82–3, 107, 128, 145, 149, 161, 164, 192–3
real 15, 42, 83, 88, 114–16, 117–18, 119, 123–4, 127–8, 143, 149–50, 192
and reality 67–8
reality of 9, 14, 17–18, 19, 70, 97, 114–24, 125–38, 191
significance of 38
theory of 4
transcendental 159–60
transfinite 118–20, 137
uncontrollable growth of 75
whole 7, 22, 131, 177–8
numerical computation 162–3
numerical fields 12, 177
numerical optimization 59–60
numerical procedures 7
numerical progression 51, 90
numerical series, theory of limits 78–9

Old Testament 18
Orestes, trial of 40
Orphics 37, 43–4
overflow 176
Ovid 48

Palamedes 24–5
Parlett, Beresford 153
Parmenides 25, 77, 97
Parmenides' thesis 114
passion 73
Peano, Giuseppe 22, 148, 151
perceptions 53, 54
Perron, Oskar 199–200
perspective, overturning of 105–6
Philolaus 26, 55, 57
philosophical expressions 36
philosophical formulas 36–42
philosophy 180–1
physical reality 151
physics 16–17, 117, 152
phýsis 45–9, 51, 59, 65–6, 70, 73, 75, 90
Pincherle, Salvatore 141
Plato 17–18, 24–5, 37, 38, 48–9, 51–2, 53–4, 63, 65–7, 72, 74, 90, 95, 97–9, 98–9, 109, 132, 140, 181
mathematical philosophy 4
Dialogues 36
Meno 51, 64
Sophist 65, 131
Epinomis 66
Platonic Good 49
Platonic vision, the 35
Plautus 192
Plotinus 68–9, 70, 131
plurality, paradox of 89–96, 114

Poincaré, Henri 85, 107, 140, 145, 146, 188
points 90–3, 114, 118, 126, 211n1
power 63
Prajāpati 32–3, 35
predictability 161
pre-established harmony 146
Proclus 26, 43–4, 61, 65, 90, 97, 107, 109
production 64–5
progression 35
ad infinitum 94
Prometheus 24
proportions 35, 39, 125
Proteus 49
Proust, Marcel 59, 122
Psellus, Michael 70
Pythagoras 25, 44–5, 45, 67
Pythagoras' theorem: 28, 61–3, 99, 113
Pythagorean numbers 55–7
Pythagoreanism 4, 41
Pythagoreans 4, 23, 26, 37, 38, 95–6, 119

quantifiers, logic of 18
quantities, growth of 6–7
Quine, W. V. 14, 15, 127, 131, 136–7

Raphson, Joseph 30, 57, 59, 205n6
rational corpus, the 107
rational diagonal, the 98, 109–10
rational numbers 22, 82–3, 107, 128, 145, 149, 161, 164, 192–3
real functions, theory of 134–5
real numbers 15, 42, 83, 88, 114–16, 117–18, 119, 123–4, 127–8, 143, 149–50, 192, 216n4
realism 16–17, 135–7
reality 3, 4, 11, 15, 17–18, 90, 94–5, 98, 108–9, 181–2, 192
and algorithms 154–6
external 38
mathematical 5
of mathematics 142
and numbers 67–8
of numbers 9, 14, 17–18, 19, 70, 97, 114–24, 191
physical 151
Platonic definition 97
of relations 13
realizability 155–6, 177
recurring relations 163–4
recursion 196–7
theory of 148–9
recursive calculus 196–7
reductionism 181
regula falsi, the 39–40
Reichenbach, H. 215–16n2
relation 17–18, 39, 47–8, 68, 99

relationality 4, 66–7
relations
  equality of 130
  Eudoxos/Euclid theory of 128–9
  reality of 13
  sequence of 104–6
religion, and mathematics 24–6
religious observance 7
Renaissance, the 30
residual error 106
Richard's paradox 147
Riemann, Bernhard 129
ritual 8
Robinson, G. 168–9
Rogers, Hartley 155–6
rounding 220–1n3
Russell, Bertrand 12–13, 13–14, 18, 125, 127, 131, 146–7
  Our Knowledge of the External World 13
  Principles of Mathematics 80–3, 84–7, 120, 120–1, 138
  theory of descriptions 136

Sachs, Abraham 41
Śankara 33
Śatapatha Brāhmana 25, 31–5
sceptical nominalism 135–6
schemas 6–7, 8
scientific discoveries 15–16
scientific theory 16
sections 64, 127, 131–2, 134, 194–5
semantic paradoxes 5
set theory 22
shapes 26, 68–9
similitude 94
Snell, Bruno 73–4
Socrates 65, 72
soul, the 37, 55, 206n11
  functions of 53–4
  growth 73–4
  and nature 73
  nature of 72–5
  passions of 73
space
  mathematical reality of 144
  and time 15, 121–2
space-time 83
spatio-temporal continuum, the 83
speed 82
spermatikòs lógos 36, 69
Spinoza, Baruch 73, 181–2
square, enlargement 29–30
square roots 59, 158, 208n4
stability 58, 213n15
Stoics 37, 69, 73
straight line–number relation 43
Strang, Gilbert 168
Stranger from Elea, the 109–10
subjective fictionality 146

Śulvasūtra 27–34, 45, 50, 56, 62, 99, 113
syllogistic theory 125
symbol-numbers 107

Taittirīya Samhitā 31
Talmud, the 21
Taylor, A. E. 52
Taylor, Brook 58
technique 64–5
temporal sequence 54
Terence 192
ternary set 92–3
Theon of Alexandria 30
Theon of Smyrna 104, 109, 110
theories 4, 8
thesis 26–7
Thibaut, George 29, 56
Thomas Aquinas 37, 138, 195
Thoth 24
Thureau-Dangin, P. 61
time, and space 15, 121–2
Toeplitz matrices 122–3, 215n12
transcendental numbers 159
transfinite numbers 118, 137
truth 12, 36–7, 38–9
Turing, Alan 5, 75, 131, 150, 156, 169, 174–5, 185, 199–200, 201n2

universal key 42
universal quantifier, the 18
universe, the 58, 93
Upanishads 33
Uranus 25

variables, limits of 82–3
variations, instability 30–1
vectors 166–8, 222n1
Vedic geometry 23
  fire altars 27–35, 36, 43, 56, 64, 99
  and Greek geometry 28
  and incommensurability 112–13
Vedic mathematics 4, 6, 7, 8, 41, 140–1
Vedic tradition 25–6
Veronese, Paolo 22
Verum et factum convertuntur 192–5
Vico, Giambattista 39, 192
Viète, François 30, 59, 59–60, 70
virtual realization 155–6
von Neumann, John 15, 147, 150, 173–4, 176, 180, 184, 185, 191
votive offerings 26–7

Web, the 198–9
Weber, Heinrich 132
Weierstrass, Karl 15, 22, 81, 83, 86, 96, 99, 105, 106, 141, 170
Weil, Simone 17, 27, 68, 74, 130, 194, 206n1, 213n10

Weyl, Hermann 88
Whitehead, A. N. 79, 83–4, 86, 117, 131, 138, 142, 168–9
whole numbers 7, 22, 131, 177–8
Wiener, Norbert 54, 122, 124
Wiener–Hopf equations 122, 123
Wilkinson, James 169, 172, 174, 224n12
world, the, engagement with 3

Xenocrates 72, 72–3

Zeno of Citium 73
Zeno of Elea 25, 76, 79–80, 89–90, 92, 100
Zeno's Paradoxes 4, 7, 26, 76–88, 89, 93, 114, 117
  of Achilles and the tortoise 77–8, 80, 86
  Aristotle's critique 78–80, 86
  and the infinite 77–80
  Russell on 80–3, 84–7
  solution 80–8
  third 78–9
  Whitehead on 83–4, 86
Zermelo, E. 147
Zeus 40